Enterprise Diagnostic Method

Organizational Performance Improvement

According to the Hoshin Kanri

Alignment and Lean Enterprise

1st Edition

GlobalSouth

Press

Robisom Damasceno Calado, Antonio Batocchio,
Felipe Araujo Calarge, Messias Borges Silva

Published in the United States by Globalsouth Press Inc™.

Published in the United States of America

For more information, please contact info@globalsouthpress.com or

go to http://www.globalsouthpress.com/

Translated by **Waleska Jungmann**
Revised and Edited by **Ligia Milla**
Book design by **Héctor Guzmán**

Enterprise Diagnostic Method

Organization Performance Improvement

Includes bibliographical references and index.

ISBN:

978-0-9899683-9-3

1.Lean Production. 2. Benchmarking.
3. Strategic Planning. 4. Lean Production.

GlobalSouth
PRESS

PREFACE

The objective of this work is to demonstrate how to diagnose and identify the strengths and weaknesses of production management in companies of different sectors, with the objective of providing aid subsidies for investment decisions, strategic information and opportunities to improve processes through the Lean Production focus. The conduct of the Enterprise Diagnostic Method (EDM), according to the Lean approach, goes through workshops for self-assessments in the company with a broad questionnaire, the applications of the Value Stream Mapping tool (VSM) or in the operations and services of the company, which allows planning actions for a better management of the team capacity and to orient Hoshin planning and Strategic Management in a more integrated and sustainable way. The EDM applications give as a result a quantitative evaluation of the maturity degree of the business organization, applying and considering the concepts related to Value Stream Mapping, Lean Production, Capacity Management, Grey System and Industrial Benchmarking. The study has as objective to create the conditions to help middle and high management decisions in short a middle terms through definition of processes, designs, performance improvement of actions in the organization according to the alignment to Strategic Management (Hoshin Kanri) and the focus and principles of Lean Production.

SUMMARY

1. INTRODUCTION

People employed in private and public enterprises carry out the transformation of products and services, and in this temporal function production is achieved through the effects of global competition. Within the strategic business management and operational strategies are developed to support its survival, increasing profitability and achieving market growth, as well as in the competitive strategic vision today there is a stronger necessity for the production function to be aligned with other central and support functions.

Priorities and objectives of operations and logistics must follow previous queues and remain sustainably aligned to the business itself. In the operations and services management several actions are taken for continuous improvement in search for minimization of configurations and changes in machinery, e.g.reduction of inventory and product standardization.

The current turbulence has accepted the necessity to understand and satisfy the needs of the customer through obtaining cost and value advantages, but there are always companies that have not implemented Total Quality Control Systems and/or Total Quality Management production systems in a more sustainable way.

1.1 EDM contributes to the transformation of the business

A production system is a requirement which becomes more important for management decisions and day to day routine management, these decisions are taken according to the general strategy of the company, which is why it is interesting to study the implications of Overall Equipment Effectiveness OEE for smaller decisions, e.g. during the deployment and release phases or the transformation of a production system which will have an impact in the whole supply chain, whether directly or indirectly.

The Enterprise Diagnostic Method (EDM) is applied to companies for adjustments and top down perspective, with a scientific methodology, where in the end of each questionnaire, the directors will be presented with a technical report to carry out the planning and strategic management.

Evaluating members in the business who participate benefit from the contribution of knowledge of the contribution that will be received and/or developed in relevant and updated topics, such as Lean Enterprise, Grey Theory System, Hoshin planning, Strategic Management, Benchmarking and Capacity Management. This knowledge is fundamental for the increase of assets in companies and to achieve in an integrated manner the economic objectives of the business.

This knowledge, developed in a participative manner with hierarchical level representatives in the studied company, contributes to the development of the Enterprise Diagnostic Method (EDM). It may also maximize the investment through conventional production management transformation with the application of Lean Enterprise.

A strategic alignment between the production system and joint company objective development is important. For this, the following questions must be answered: How can we change the conventional fabrication system to comply with strategic company goals? How can we identify project opportunities for improvement that are aligned to organizational levels?.

Companies are in different maturity levels. The application of methods and tools depend of the structure according to principals and adjusting to the reality of each business. There are some sustainable actions and programs that may not be aligned to strategic decisions, therefore, with a diagnose and alignment of objectives and directions and the adequate election of tool applications and efficient manufacturing methods, a transformation of the production system can be achieved, with considerable improvements and a cultural change.

When a person is not in good health, it is natural that they may call on a doctor's office for assistance in a clinic or hospital. After going through reception and identifying them, the patient is treated by a doctor. The doctor listens to their ailments and starts building a treatment plan. In the health area, the diagnostic is carefully built, since a human life depends on it, in which the added value is high and unmeasurable, before, during and after the treatment. The doctor also seeks to monitor improvements to verify if the patient is in perfect conditions. If companies are similar, the diagnostic implements projects and improvement actions that are always accompanied by activities through the addition of strategic management to assure business sustainability.

We present in this book a method for diagnosing enterprise performance, to be able to systematically see and propose leverage for integrated projects and actions aligned to Hoshin Kanri and Lean Enterprise. We suggest management and measuring of performance of the integrated system based on the top down business strategy and implementing an operational strategy vertical to the hierarchical structure, based in cases of organizational reality, specific to companies of different sectors.

The application of Hoshin Kanri is the development of a panning method for strategic management in order to support the deployment and/or transformation of a conventional production system, but it is necessary to know that the changes and improvements of the manufacturing and service fabrication performance depends on top down decisions and their strategic management.

The Lean Enterprise concepts integrated with the diagnostic method help in the resolution of problems and implementation of sustainable improvements. The contribution of the European detention order is described in this book and offers an opportunity identification system for better and simple results in the company and evaluates the degree of maturity of businesses, meaning, it is the road for the company to achieve better competitively rates, improving quality, productivity and finally to know how long it will take them to become a world class business.

1.2 Financial performance of the Toyota production system

In Japan, by the end of the XIX century, Sakichi Toyoda invented the first electronic weaving machine in Japan in 1918 and founded Toyoda Spinning and Weaving Company. In 1924 with the help of his son, Kiichiro Toyoda, he created the automatic loom. Kichiro then visited Detroit, United States, and became an automobile production enthusiast. When he returned to Japan, he founded the Toyoda group with an automotive arm and produced de model AA car in the division, followed by the creation of the Toyota Motor Corporation in 1937.

Toyota has been in the latter years among the ten largest companies according to the world ranking of the 500 largest electronic businesses in Fortune magazine, the classification is based upon income of millions of dollars. This rating also shows that since year 2008 (Table 1) Toyota started to occupy the first place in the ranking of automobile companies, positioning itself as the best company in the industry.

During 50 consecutive years Toyota invested in people and created its own culture which translated into a continuous growth and earning their market share. The company is a synonym for customer oriented service, reliability, , environmental responsibility and world class quality amidst difficulties within the recession and retirement crisis, in which it had a loss of over $4 billion in 2008. They were accused of a late and inadequate reaction to the crisis, by denying or even hiding the facts and supporting security benefits, which carried the aforementioned built image.

The next table shows Toyota in 2005 in third place and in 2007 in second place, reaching the top and taking the first place in the ranking among competitors. In the last decade, Toyota remained within the ranking of the largest electronics companies out of the world's 500 largest ones featured on Fortune Magazine. Comparing results from 2014 to those in 2005, it seems that their millionaire income improved 48%.

Year	Global Ranking	Automobile Ranking	Revenue (in Millions $)
2005	7	3	172,616.30
2006	8	3	185,805.00
2007	6	2	204,746.40
2008	5	1	230,201.00
2009	10	1	204,352.00
2010	5	1	204,106.00
2011	8	1	221,760.20
2012	10	1	235,364.00
2013	8	1	265,701.80
2014	9	2	256,454.00

Table 1. Toyota Motors Leadership in the world automobile ranking

In the following graph, in which 2014 is compared to 1995, we find that the number of work positions in Toyota grew 202% in the last nineteen years, from 110,534 to 333,498 employees, in comparison to their competitors. Toyota has developed the capacity of sailing through troubled waters and has a higher capacity today to face the crisis. The double crisis was in fact more of an opportunity to motivate all the Toyota team members and partners towards a vision for improved quality and security to a never before seen level in the automobile industry, the sustainability of Toyota lies in the fact that its cultural foundation turns crisis into opportunities.

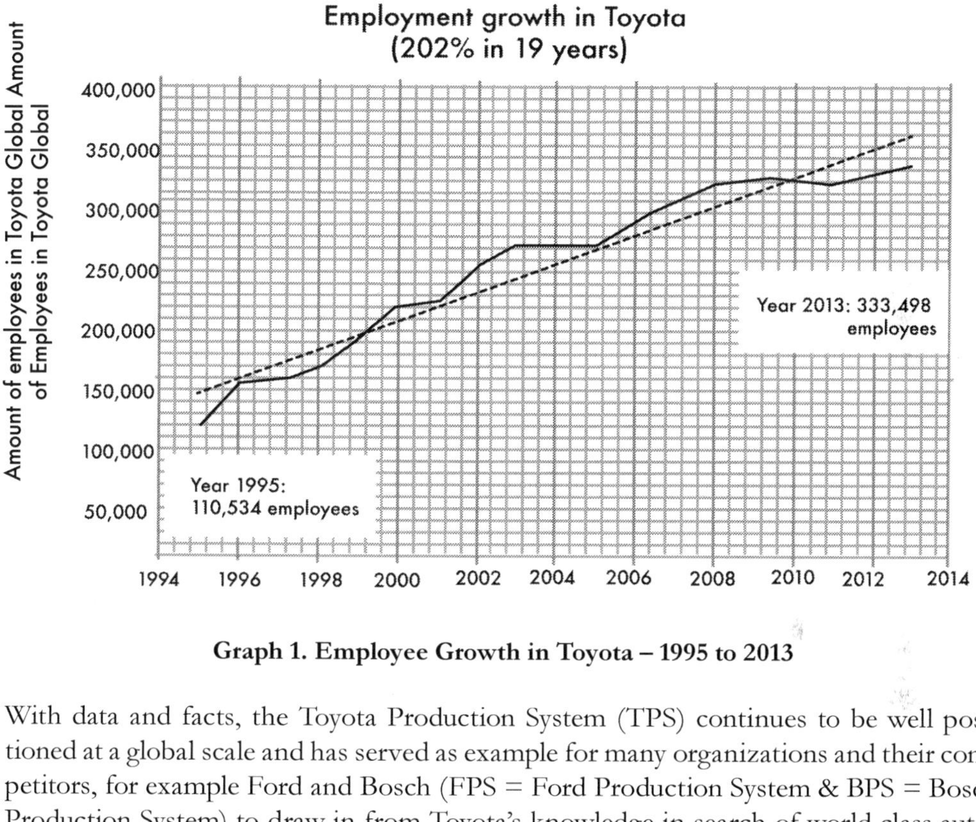

Graph 1. Employee Growth in Toyota – 1995 to 2013

With data and facts, the Toyota Production System (TPS) continues to be well positioned at a global scale and has served as example for many organizations and their competitors, for example Ford and Bosch (FPS = Ford Production System & BPS = Bosch Production System) to draw in from Toyota's knowledge in search of world class automobile leadership. Another example is the production system of BSH GmbH (BSH), which is a German appliance maker, was founded in 1967 and is a consortium between Siemens AG, Stuttgart and Robert Bosch GmbH, Berlin and Munich. One of the authors, Dr. Robisom D. Calado, worked with technical processes as a Process Engineer and Head of Industrial Engineering, where he coordinated improvement programs such as Lean Manufacturing from 1998 to 2009. Dr. Calado has accrued 25 years' experience in the industry, in Industrial Engineering for middle and large enterprises.

Lean Production defies production practices, reviews and vitalizes production and quality concepts in the supply chain. According to investigators Womack, Jones and Roos in their original publishing in "The machine that changed the Word" regarding the Japanese automobile industry showing the world its superiority in fabrication, Toyota may be more competitive with their production system through their performance that obliges their competitors to learn from stakeholders, even from the most modest operators.

1.3 Book content

This book approaches EDM according to the Lean Production guidelines. It is divided in seven chapters; the first introductory chapter had a presentation of the main objectives of this book, as well as its contribution.

Chapter 2: presents a revision of the literature on the approach, methods and Lean tools, which provides the comprehension of continuous improvement and stimulating the better management of team capacity.

Chapter 3: Relevant concepts that are the basis of investigation will be described. It touches relevant subjects, such as comparative assessment, strategic management according to the Hoshin Kanri approach and the Grey System as a selection and assessment indicator method.

Chapter 4: presents a method proposal to diagnose the potential and opportunities and improvements of integration with Hoshin Planning and strategic management to transform and apply Lean, the operations, logistic and operations system.

Chapter 5: In order to contribute and validate the EDM budget, this chapter shows the application of the Enterprise Diagnostic Method in a mechanized company in the industry.

Chapter 6: presents a discussion regarding the obtained results in the application of the Enterprise Diagnostic Method in four different companies.

Chapter 7: in the last chapter the main conclusions, considerations and recommendations for future works are presented, based in market studies.

2. CONCEPTS AND TERMINOLOGY OF LEAN PRINCIPLES

2. 1 Competitive advantage

Christopher M. (2007), with his commercialization and logistic vision, believes that success in the market is related to competitive advantage and the three Cs: Company, Competitors and Customers. An advantage must be provided to the customer at a price they are willing to pay, meaning, an advantage of cost and value, in which the value advantage is reliable and sensible, with personalized services. The cost advantage is having synchronized supplies, asset rotation and capacity or ability management, using the work force correctly. According to Christopher support activities must be organized within a matrix and sequentially in order to achieve the desired margin of benefit for business continuity (Figure 1).

Logistics and supply chain management may provide a competitive advantage over their competitors in terms of customer preference through customers and through an improved management of logistics and the supply chain.

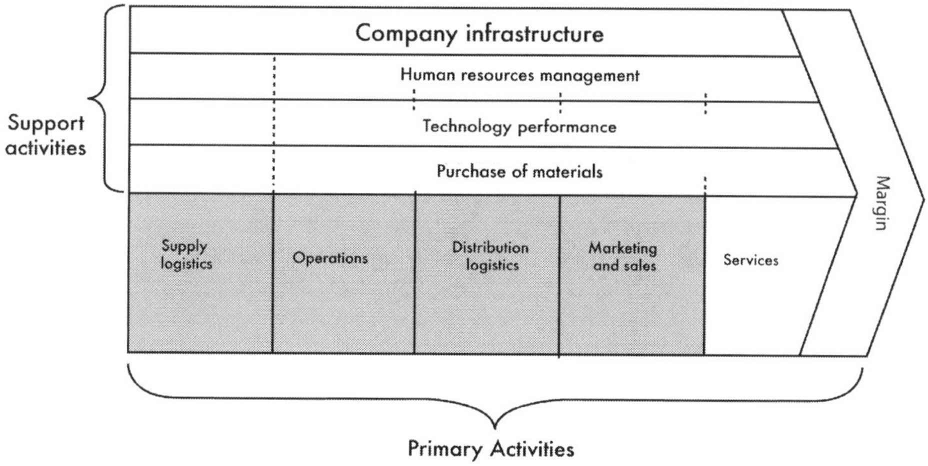

Figure 1 The Supply Chain (PORTER, 1985 aped CHRISTOPHER, 2007)

Today's competitive advantage is effective time management, as per Daugherty and Pittman (1995), strategies based on time of distribution must be used, from production until the point of sale, when the delivery does not supply the needs of customers, opportunities are lost and without the fast fabrication process, the advantages are lessened, since there is a void that has been left to be filled by another competitor. The company needs to achieve good response levels throughout the supply chain with higher flexibility and agility (REICHHART; HOLWEG, 2007).

In the pushed production system (figure 2), which classifies itself as traditional materials moving to the next stage as soon as they are processed, in general larger lots with long lead times. The administration of the transformation of the raw material is not an easy task.. Management is induced to act in a corrective manner. Searching for the use of fabrication capacity and higher production volume and production excesses is when the objectives can be accomplished. Generally, there are difficulties to comply with that mix, meaning, serving the customer. The problems are not as visible and they are not easy to resolve. Interruptions in the flow are more constant and the final quality, as well as the costs is not as competitive in a global scale.

In the pull production system, materials are moved only when the next step is called, generally in small lots with shorter lead times than the traditional system, which makes the Value Stream of the supply chain management easier. The flow manager maps the chain and strives for perfection, attempting to act in a preventive way. Their biggest concern is producing only what the customer needs, with guaranteed quality, avoiding interruptions in the production process, as shown in figures 2 and 3 (SLACK et al., 2009).

It is understood that fabrication comes from a project with operational requirements to serve customers based on parameters and its structure is not rigid but evolving continuously. It is important to explain this as follows in these five principles for Toyota Production System; Value; Value Stream; Flow; Pulling production and Perfection.

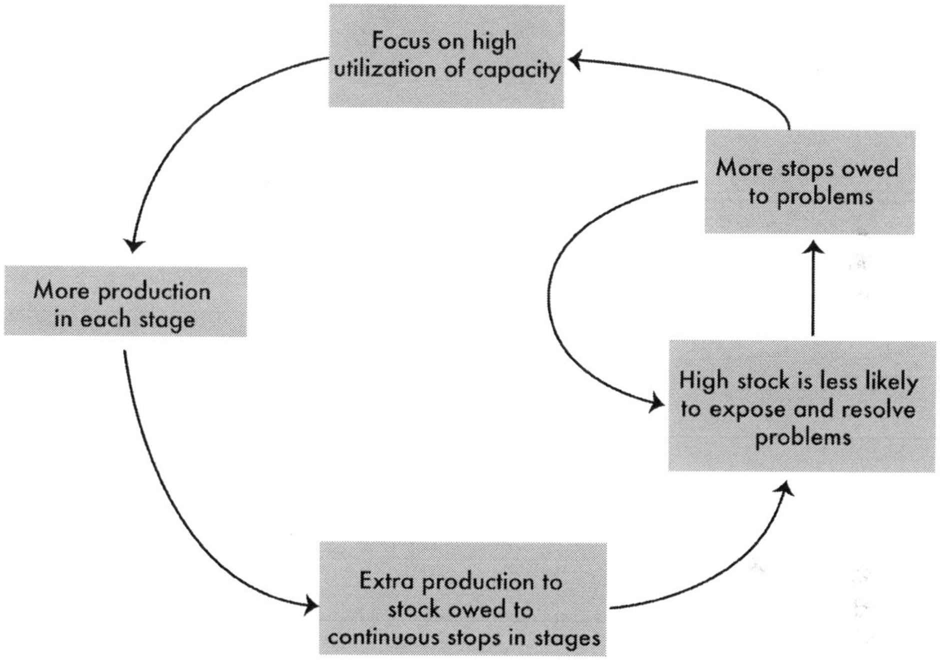

Figure 2 Push Production Analogy (SLACK et al., 2009)

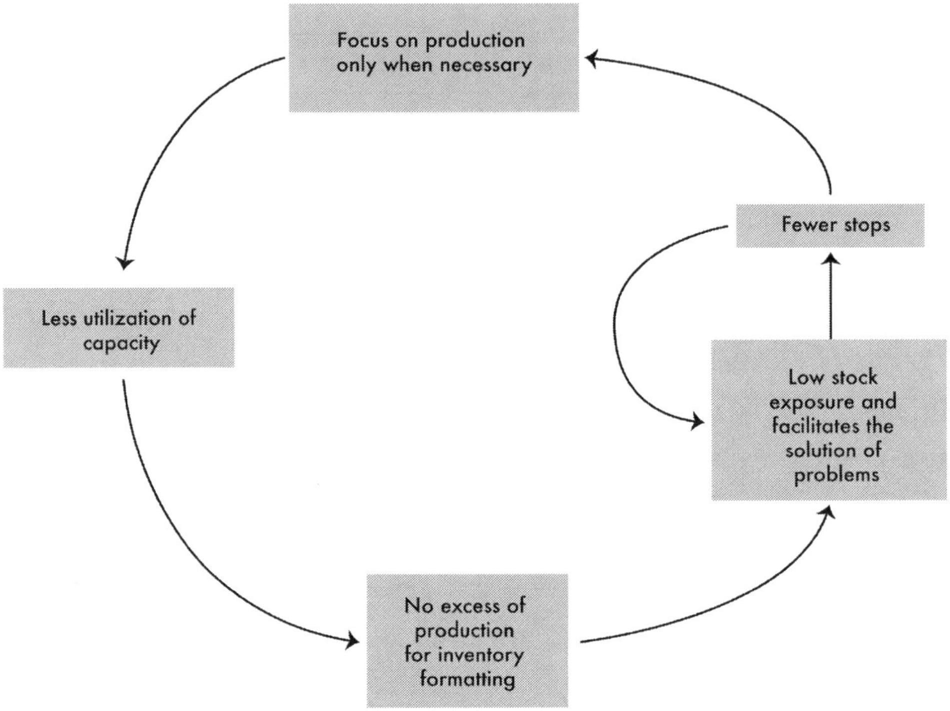

Figure 3 Pull Production Analogy (SLACK et al., 2009)

2.2 Toyota's model structure

Ghinato (1995) in figure 4 shows, according to his investigation, the use and alignment of tools and methods with the Toyota Production System model and Womack et al. (1998), places no structure to achieve lean thinking, but suggests Hoshin planning, which is a tool for strategic decision making for the executive team in a business. This tool focuses on resources in the critical initiatives required to reach business objectives of the company. Using matrix diagrams similar to those used in the quality function deployment (QFD), selecting up to three or five key objectives and ignoring all the rest. The selected objectives are translated in specific projects and are deployed at the implementation level in the company.

Strategic deployment unifies and aligns resources, establishes measurable objectives clearly, measurable in relation to fundamental objectives which are measured regularly. In the book "Lean Thinking" Womack and Jones (1998) claim that lean thinking have better results. The five year commitment is divided in four phases (Chart 1), to achieve the lean structure of the Toyota Production System, as well as the time required for the suggested implementation of the main steps for the lean production system.

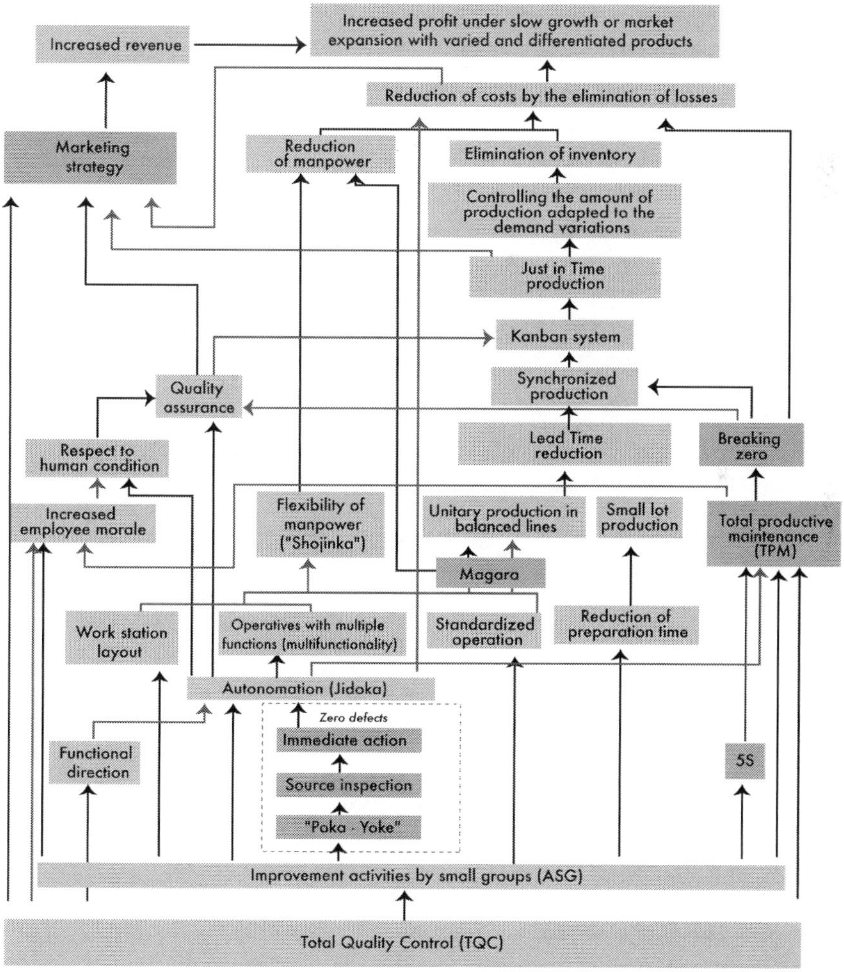

Figure 4 Toyota Model Structure (GHINATO,1995)

Chart 1 Timeline for Lean Thinking Phases (WOMACK; JONES, 1998)

Phase	Specific Stages	Term
Beginning of the process	Find a change agent Learn Lean Thinking techniques Find leverage Map Value Streams Start flow kaizen Expand the scope	Six months
Create a new organization	Reorganize through product family Create a lean function Develop an excess staff policy Develop a growth strategy Eliminate obstacles Install a perfection mentality	Six months to two years
Install Enterprise Systems	Introduce Lean Accounting Implement transparency Start deployment policy Introduce Lean Thinking learning Find the right size tools	Three to four years
Complete Transformation	Apply this measures to your providers and customers Develop an integral strategy	End of five years

2.3 Lean approach principles

In engineering and management it is understood that a manufacturing born of a project with functional needs to serve customers is based on design parameters and its structure is not rigid, and the five principles of the Lean Thinking and Toyota Production Systems , can be identified. According to Womack and Jones (1998), the five principles considered as a base for lean production in the Toyota Production System, but later came to light the sixth Lean principle related to people(Oppenheim,2011), the principles are:

Principle 1 VALUE: capacity offered to a customer at the correct time for an adequate price, according to what was defined by the customer.

Principle 2 MAP THE VALUE STREAM: specific activities to design order and offer a specific product, from its concept until the launch of the order and its delivery, and the raw materials to the customer's hands.

Principle 3 FLOW: progressive realization of tasks throughout the Value Stream for a product to go through from its concept to its launch, from the request until delivery, and the raw materials to the customers hands without wasted flows.

Principle 4 PULL PRODUCTION: production system with delivery instructions for activities derived from planning activities in which nothing is produced by the raw material supplier without the need of being pointed out by the customer.

Principle 5 PERFECTION: total elimination of any activity that consumes resources but does not create value for all the activities in the time of a Value Stream to create value.

Principle 6 – RESPECT FOR PEOPLE: a Lean Enterprise is an organization that recognizes that its people are the most important resource and the one that adopts high performance work practices. In Lean enterprise , people are not afraid to identify problems and imperfections with honesty and openly in real time, brainstorm about root causes and corrective actions without fear, and plan effective solutions together by consensus to prevent the problem from occurring again.]

In the latest decades, high business executives have adopted improvement projects in an effort to transform companies to compete successfully in the future. Examples of improvement initiatives are: Total Quality Management; Production of Distribution Systems (just-in-time); Competition based on time; Lean Business Production; Creation of Customer Focused Organizations; Cost Management based on Activities; Employee Empowerment and Reengineering.

Improvements require important changes and the objectives of these programs are not improvements that would increase the enterprise´s survival however they do allow a performance that enables success in the information and knowledge era (KAPLAN; NORTON, 1997).

To achieve excellence in fabrication, principles for lean production can be implemented and the objectives show what the desired achievements are. The methods and tools, according to the lean production approach, are derived from principles for the desired objectives. These methods describe the processes to implement, in the way in which the objectives are to be achieved and they are represented by the tools. **Figure 5** describes the five basic principles of an operations and logistics system, adapted to a refrigerator company researched in point 2.9 (Implementation Evolution of Lean Methods and Tools), whose base is the normalization and sustainability. The figure is also based in the definition of the five principles of organizational lean thinking and the Toyota Production System (WOMACK; JONES, 1998).

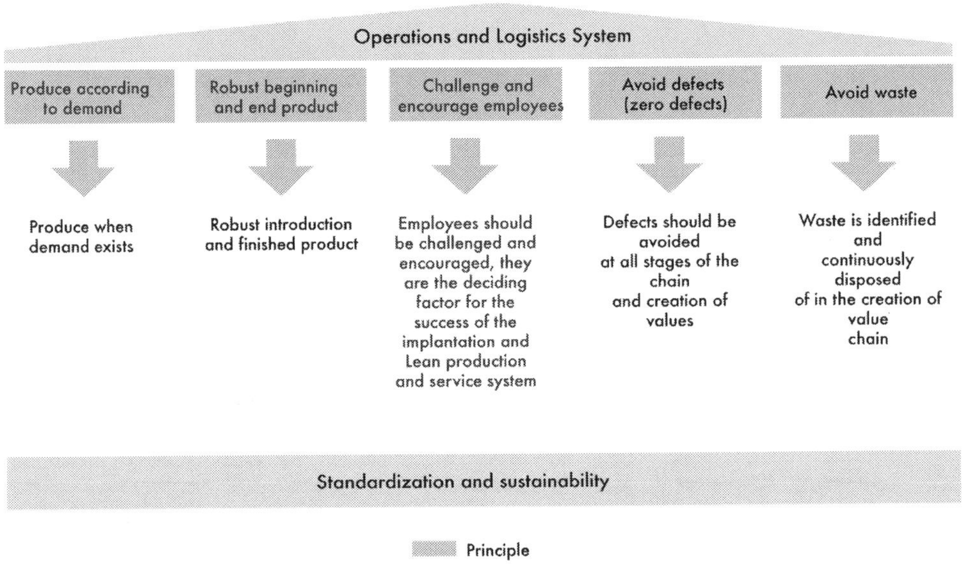

Figure 5 Basic Principles For Operations and Logistic Systems

The capacities of more fabrication processes are widely synchronized and aligned to the methods and tools. **Figure 6**, based on the approach of Slack et al. (2008) describes the objective (lean synchronization) and the approach to overcome barriers to implement lean synchronization methods and diverse techniques that may be used to help suppress losses.

Miyake (2008) presents a model with the basic tools in accordance to Lean principles (figure 8), but Slack et al. (2008), on figure 6, deals with principles and approaches. The principles are different between the surveyed multinational companies such as: Bosch, BSH, Eaton, and Alcoa. The investigator has followed the evolution of the Lean application within these companies for the past few years and the categories that each company has established as their principles were chosen according to their realities and needs. The difference between them, related to the principles is due to each company's needs, the discussion and definition that correspond with strategic objectives in a personalized manner. In figure 7 the Miyake model (2008) is presented, which is a proposal for Lean steps in this alignment to principles and tools.

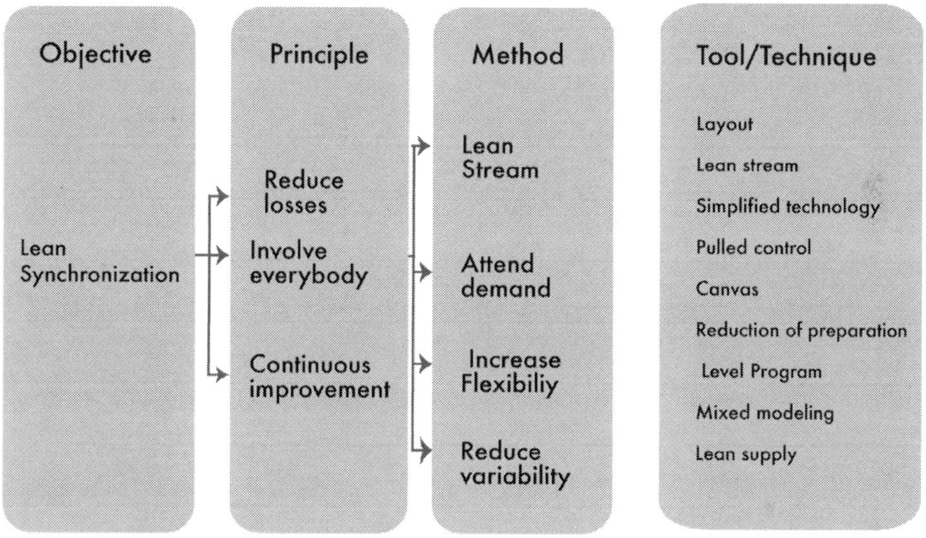

Figure 6 Lean Synchronization (adapted from SLACK et al., 2008)

Figure 8 describes an implementation process of a lean production system and the transformation from a conventional production system to a leaner configuration, the approach is known as Lean Principles, and it is also used to overcome adversities and obstacles, making the implantation of Lean Thinking easier. The tools presented are described by Slack, in some cases, as methods, even though both authors objective is to eliminate losses and higher added value for customers (MIYAKE, 2008).

According to Feld (2000) and Satolo et al. (2006) the methods and techniques that are generally used in the application of lean production (figure 9), with the objective of battling residue sources and process improvements, can be grouped in five main categories, which are:

Production Flow – methods and techniques related to physical exchanges, product development procedures, and definition of the required coverage norms.

Organization and culture – it refers to matters related to individuals, learning, communication and shared values within the organization.

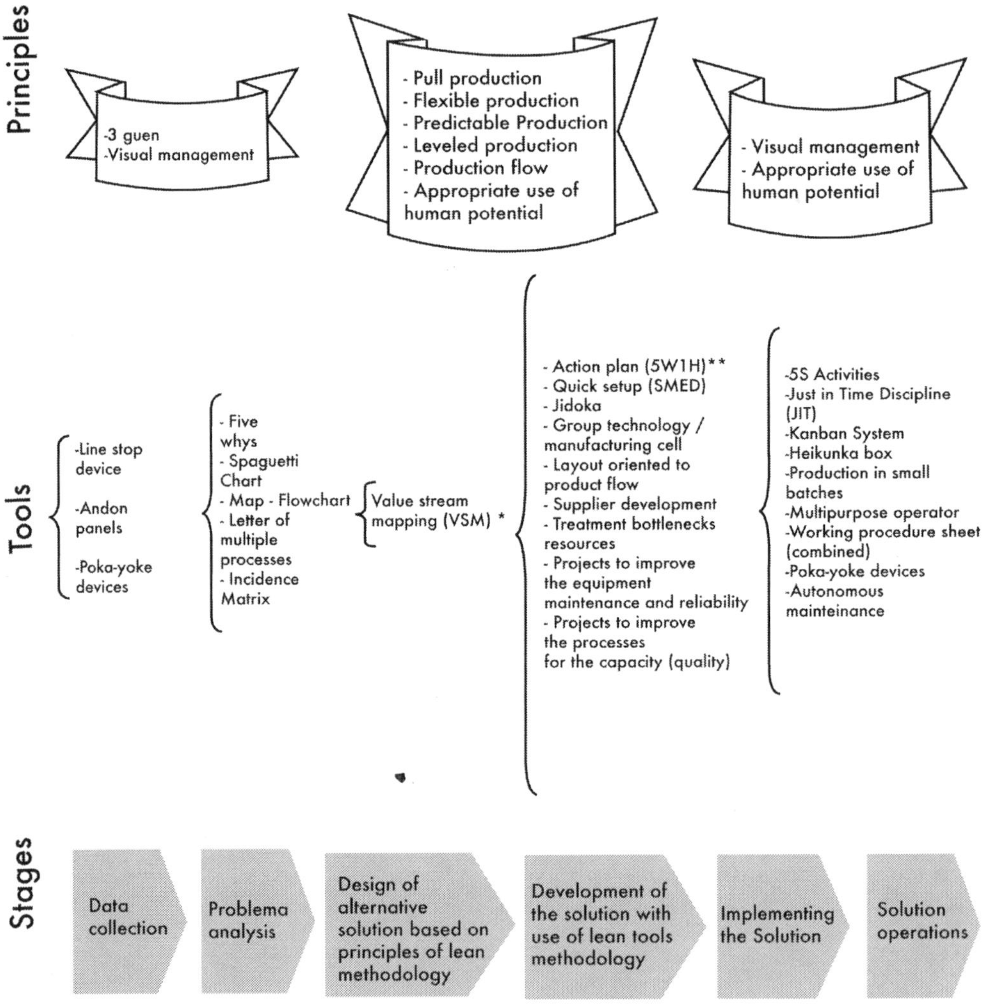

Figure 7 Lean Model Principles and Tools in Steps of an Improvement Project.

(MIYAKE, 2008)

Process control– tackles methods and techniques related to monitoring, control improvement stabilization in the production process.

Metrics – include methods and techniques that measure performance, improvement goals and rewarding performance of work teams and employees in the organization.

Logistics – norms and procedures, techniques and planning methods and internal and external material control related to organizational flow.

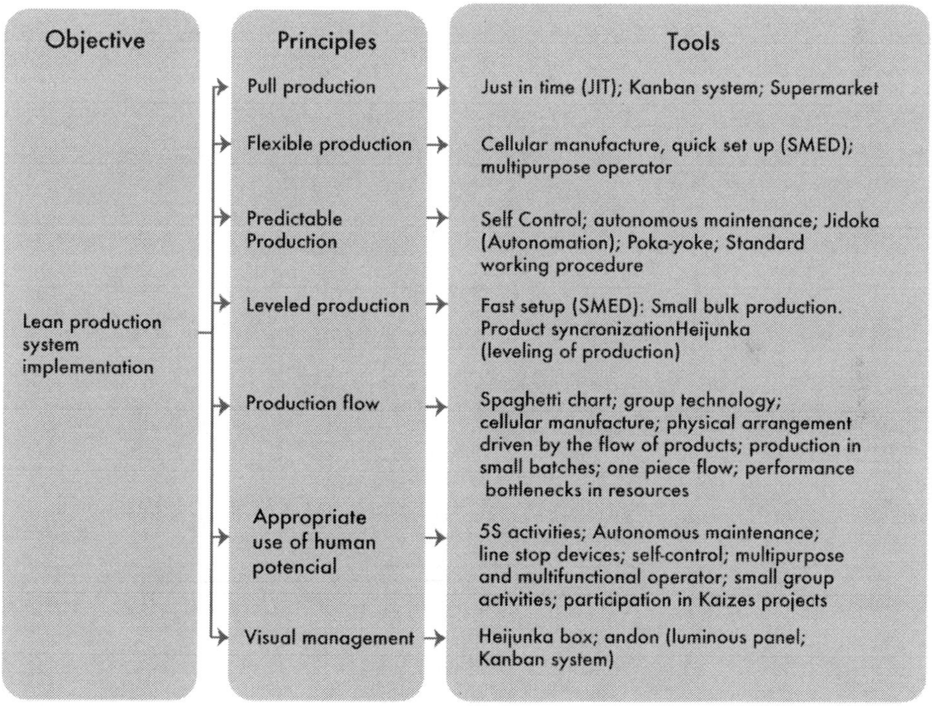

Figure 8 Associated Lean Model Principles and Tools (adapted from MIYAKE, 2008)

The exposition of principles and alignment approaches, as well as Lean methods and tools cited by Slack et al. (2008), according to figures 6, 8 and 9 may provide a better vision of the differences and pertinence of the basic instruments in each author's point of view. These examples indicate that there is no single norm to implement the Lean tools and methods.

The following topics will contain the adjustments, even though in some cases there are differences, the basic tools and methods which are used in the Lean implementations in companies and services (MURMAN et al., 2002) and the structure of the basic principles for operations and system logistics will be adopted (figure 5), but as part of the research and development of investigators in a multinational refrigerator fabricating business in the past five years.

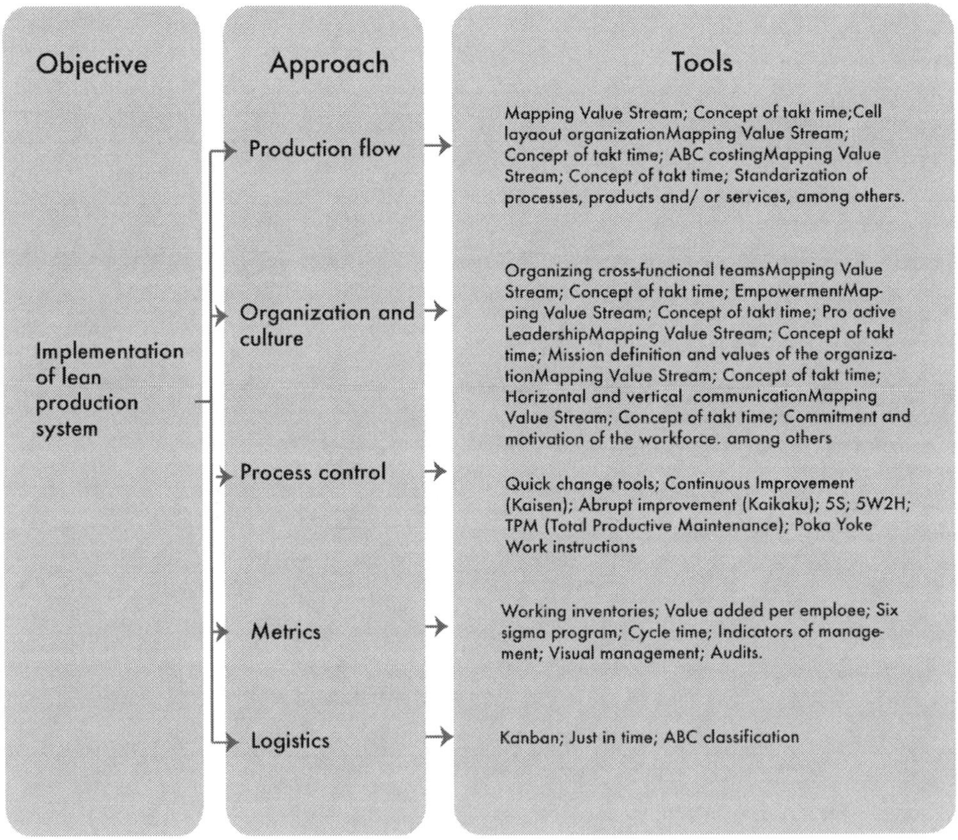

Figure 9 Methods and Techniques used in an Lean Production System (Adapted from SATOLO et al., 2006)

Section 2.4 is a revision of literature regarding the concepts of the basic principles of the operations and system logistics (figure 5) and their methods and tools, which will include their bibliographical references.

2.4 Producing only according to demand

The principle of producing only according to demand means the optimization through the added Value Stream and orientation to the customer's demand. Producing only when there is demand means producing by the demand of customers and delivering in time with shorter time spans. This customer oriented effort may increase their satisfaction; it avoids overstocking and generates simple and transparent processes. Therefore, the challenge for businesses is having the right product available, in the right time, place, amount and quality desired by the customer at the minimum cost.

To produce only when there is a consideration of customers demand requires the application of knowledge and methods such as: the pull system, takt time and supply management (figure 10). It is worth mentioning that the presented principles are defined according to each business strategy: This study references a manufacturing company which is cited in chapter three.

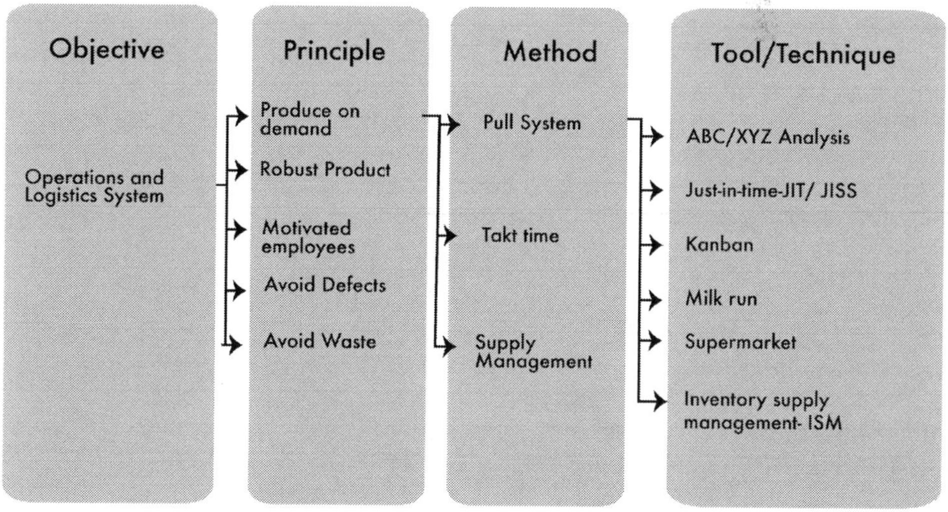

Figure 10 Pull system tools in manufacturing

Consider, as an example the usage of an appropriate supply chain and external clients, wholesalers, retailers, and subsidiaries, as well as internal clients that may have distribution centers and storehouses that belong or are under control and management of the company.

2.4.1 Pull system

The pull system (figure 11) is a system to control processes, fabrication orders are pulled to fabrication. The impulse for the execution of fabrication orders starts with the end of the logistics chain (the customer). This means that the order and/or real customer commands the consumption areas (DAILEY, 2003).

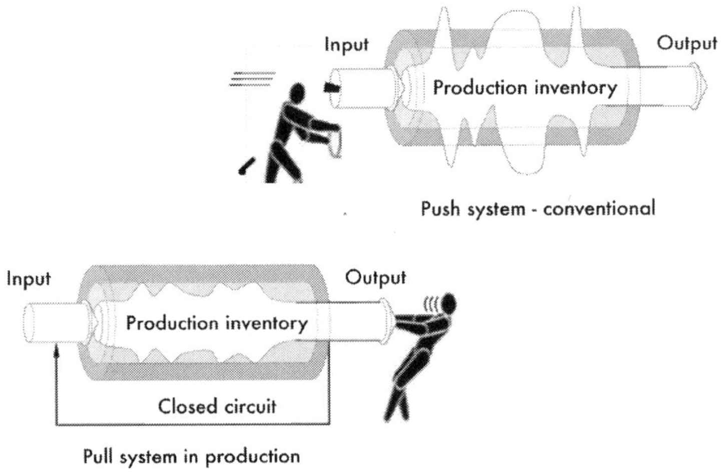

Figure 11 Analogy of the Pull and Push System (SLACK et al., 2009)

The pull system objective produces only a part, for as long as possible, when there are needs for consumption, during a short period of time it is the release of an action between two stages in the process. The actions result in the reduction of inventory, travel time reduction, increase of performance in delivery and improvements in the planning process, control, programming and production.

The pull system gives an advantage to the process chain (customer – production – provider), since all the steps in the process are connected to one another through the pull system, given that the customer´s requests trigger the production in a flow that created an adjustment point in the total process. The owners of the production control, which is planned, are replaced by an auto controlled processes. Another advantage is that sales forecasts are used for planning and demand during the middle and long terms of capacity.

The ABC/XYZ analysis is the combination of the analysis of value /amount (ABC) and prediction behavior/consumption (XYZ), the mix of parts/mix of products. So they are the base for the classification of materials/types of supply products/types of service (SLACK et al., 2009).

The types of supply are understood as standardized processes for the supply of materials, raw materials and provider hiring pieces for production. The types of services are understood as standard supply processes, finished products, in other words the production for end customers. The ABC/XYZ analysis is used as a base for selection and classification of types of provisions and services. For each class, among others, the following points are defined: point of delivery and stock types (examples: storehouse, supermarket, production line, etc.), the supply procedures (for example, the milk run system), and control methods such as kanban and orders.

Materials and products are periodically controlled no less than quarterly, and classified in the appropriate offer analysis types according to ABC/XYZ analysis, the same materials and products are controlled as a determined method, if necessary, stock and available if they meet the previous requirements. The parameters are reviewed continuously and possibly adjusted when there is a planning system of corporate resources (ERP – Enterprise Resource Planning). ABC/XYZ analysis is executed automatically through a tool and is used in all products, semi-finished parts and raw materials.

Just in Time (JIT) is a concept for the provision of materials for a particular delivery. JIT means that the correct materials in the correct time, in the adequate amount are delivered to the correct place. This is based upon specific customer orders. Bearing these requirements in mind, the management is executed as a generator (OHNO, 1997).

Just in Sequence (JIS) means using the correct arrangement. This is an additional development of JIT, aside from the JIT requirements; the provider must guarantee that the supply sequence is respected according to the customer's order. This tool helps reduce delivery times, reducing inventory and spaces, the reduction to the minimum movements of materials and handling efforts and a higher transparency of processes.

The provider and the services of the provider and/or logistics use JIT/JIS for adequate material groups. The capacity of the provider for JIT/JIS is always developed with the intention of defining the point of delivery, for example, the JIT provider's delivery frequency /JIS is at least once per shift, meaning the maximum amount of values for a single lap. All parameters, such as surface, organization, techniques, JIT/JIS contracts, the associations are clear, documented and available and there are no quality inspections done in the entry of merchandise in the company. A proper condition is where the provider and regular conditions have a consensus to guarantee the required quality (JACOBS, 2009).

The Kanban, defined as card in Japanese, is a tool that visually controls and manages the flow of information and regulates the supply of materials between processes. These are signs for the internal or external client, consumption of materials or product. The internal or external provider is activated to go back to offer his produce once again. When the customer and provider are in communication and act according to this procedure, no external interventions are required. This is an auto regulated – controlled circle (CALADO et al., 2005).

This is a tool for the application of the pull principle and the control to simplify the flow of materials through transparency to make tools and inventories available. It avoids overproduction and defective pieces, the relationships between provider and customer are visually verified and if there are problems in the process, they will be used to be applied for continuous improvement. The kanban control is applied and shown according to the correct material groups. Provider/Customer control circles are defined and established in supply points (consumptions sites). The standard information is contained in the kanban card: Element name, material number, amount by charger, source, bar code image (optional), customer number (optional), spare time, observation, total kanban number (MURMAN et al., 2002).

The milk run is derived of direct supply, similar to an old milk deliver car; a milk man delivers in a predefined time in a specific route and is in charge of replacing empty bottles for full ones. In the production process, through milk runs the materials supply is done in an established route within preset times in fixed amounts and compositions.

The milk run reduces moving people and minimizes costs for the supply of small amount or materials. It has the advantage of being able to implement an efficient supply with kanban. It avoids the control systems for complex transportation and specialized systems, as an example of the forklift control system (CHOPRA; MEINDL, 2003).

This system provides the regularization of supply materials, for example, a smaller route cycle can be established for two hours, in which the milk run (figure 12) makes several stops; stopovers are seen by the customer; there is a direct delivery to the place of consumption, packaging becomes standardized; and there is an integrated way to manage empty bottles.

Supermarkets represent buffers (understood as a small stock) of material between work stations, in the areas where there are difficulties to maintain a continuous flow of material. Supermarkets keep minimum and maximum amounts of raw materials, parts and construction groups for their respective processes, if the minimum amounts are not met then new ones are produced from the previous process until reaching the regular and pre-set values (OHNO, 1997).

Figure 12 Motorized vehicle for internal milk run

While the customer is in the supermarket, they take part of the process which is necessary to activate the chain for shelves and spare pieces, using the salesman, meaning the previous process. Supermarkets operate according to the FIFO principle (First In, First Out). This is just a way to aid the implementation of kanban and must be seen as an intermediate solution in the production flow path which is continuous and driven by the customer.

Provider processes are uncoupled from customer demand, for example, processes related to production lot sizes in a continuous flow. Assuring the availability of finished or semi-finished products but avoiding overproduction. It eliminates the manufacturing based on predictions that are replaced by the consumption control, in search to comply with for the pull principle.

A Vendor Managed Inventory (VMI) is where the provider assumes responsibility of the sold inventory related to the existing products with the customer: Complying with certain rules, the customer's vendor transfers their stock management to the it's materials and pieces vendor. Additionally, the provider will have access to existing inventory data in which they participate to the customer, for which there is an execution of supplies with independence and responsibility (SIMCHI-LEVI et al., 2010).

This tool that grants the provider 100% responsibility of availability of materials has been highly recommended by authors as a way of improving cooperation between companies and provides benefits such as: assuring availability of components; increases speed of response; reduces costs and order compliance process interfaces; increases transparency within supply relationships; building an association based in the provider's participation easing planning previous of the supply chain through the data of reliable inventory, consumption and demand.

2.4.2 Customer Takt

The definition of customer takt is also known by businesses as takt time is the average period of time in which the customer needs a part, in function of the available time of production.

$$\text{Customer takt} = \frac{\text{Available production time by time unit} \quad \text{(sec)}}{\text{Average customer demand per time unit} \quad \text{(piece)}}$$

Customer takt, according to Murman et al. (2002) may be subject to fluctuations and must be controlled periodically; its objective is to synchronize production with the customer's demand; avoiding the interruption of supplies, using minimum inventories; actively reducing waste, such as overproduction, transportation and inventory; satisfying customer expectations, providing the adequate amount in the right timeframe. Thereby fabrication capacities can be planned transparently, reducing the whiplash effect to a minimum, and therefore, manufacturers have learned the leveling of production and have delimitated the system to a single step, a rapid exchange and the chain of value (figure 13).

It takes place in the work station and once it is ready, it is sent to the next stage in the process, it is always one shipment at a time, work positions are designed to preferably have at the same time a transfer lot of an equal size. This instrument is called One-piece-flow (SHINGO, 2009), which represents a continuous flow in its purest form (continuous flow production); meaning that a piece is produced and sent one at a time.

The direct connection to the customer's demand may provoke great fluctuations in production, for which a production leveling tool will be obtained if the productions orders have been scheduled as a standardized and recurring program and the calendar is constantly complied with. Additionally the production must respect the customer takt during a period of time, is required by smaller lots and fixed horizons required for production.

The takt is used in the production of predefined small stock levels; the production is uncoupled with customer demands. So it is possible to reduce the impact of demand flows, whiplash effects already described. Among the advantages, excessive fluctuations can be avoided for production volumes, reducing lead times and customer response times, reducing inventory of production pieces and finished product inventory and increasing productivity by avoiding unplanned changes in the production plan (DAILEY, 2003).

Only the pull process is controlled through a production plan, the compliance of the production plan is measured by the number and sequence produced and the EPEI index. Every part every interval (DENNIS, 2008). As a defined pull process, the production takes place within a determined and constant leveling period. The lead time for replacements tends to be optimized through minimized setups and optimized lot sizes, pointing to differences between standard products and taking exotic productions into account.

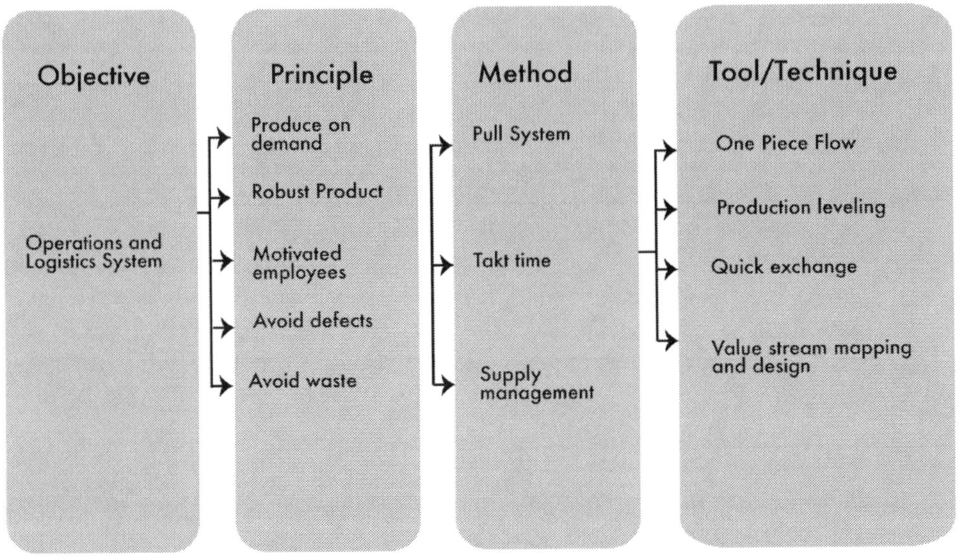

Figure 13 Tools for o takt time in manufacture

Quick changes, also known as SMED (Single Minute Exchange of Die) are a standardized process to reduce setup times by activity. It is a prerequisite for the consecution of smaller lots. The production and supply chain are favored by the tool for standardization of change procedures, reduction of exchange times, increase of exchange type, reduction of lot sizes, reduction of lead times for replacement and reduction of small populations of actions and stores (SHINGO, 1996).

According to Rother and Shook (1998), the importance of the Value Stream mapping tool is the possibility to see the flow, the visibility makes decisions easier, takes advantage of lean production concepts, it is useful as an implementation map and impulses, makes the flow of materials and information visible: It is a qualitative tool, where a first measure is done before and after reaching the desired goal.

The Value Stream Mapping tool (VSM), according to a case study from Wee and Wu (2009) has helped many manufacturing companies with waste, to identify in the application of Lean Production and continuous improvement in the long term in the case of the Lean Supply Chain (LSC), for example, Ford Motor Company in Chungli, Taiwan. has already implemented Lean Production since 1990, based on the Ford Production System (FPS) and in the case study by Wee and Wu the application of VSM is shown sharing the problem resolution method from the PDCA cycle (Figure 14) to give an industrial and development view, the four steps of PDCA for continuous improvement from the Lean Supply Chain are used.

The Lean Production approach according to Abdulmalek and Rajgopal (2007) was applied with a higher frequency in the discrete fabrication than in continuous process industries. Slack (2009) describes that the production processes for the production of goods are discrete or continuous, for example, the discrete processes for ship processes, architectonic processes or race cars, clothes by orders, furniture or plane restauration by jobbing are and by lot or mass production are cars, pizzas, ice cream or CD productions (Distribution Center). Companies such as CPFL, Petrobras, of continuous produce processes, for example, electric energy, oil refineries, chemical processes and even beer, already use many of the production practices as well as modern quality management in the supply chain.

We understand flow values as all activities that create value necessary for the fabrication of a product, from raw materials to the completion of a customer's offer. The current state of VSM is the existing chain in a product of values, the entry of merchandises until the final delivery (real situation). The future state of VSM allows the design of a future situation with the inclusion of improvements in required processes. VSM can already be used in new production facilities.

The visualization of a complete Value Stream and unraveling of the ideal Value Stream. Relation between information and material flows are represented and easy to understand, for example identify and reduce existing waste in a flow of value, defining and implementing specific optimization measures (for example: lead time, inventory, synchronization of upstream and downstream processes in a Value Stream). The analysis and design of all values in the Value Stream are done in regular intervals with factory flow improvements. The Value Streams for all product families are evaluated by mapping the flow values and optimizing with the design flow values (ROTHER; SHOOK, 1998).

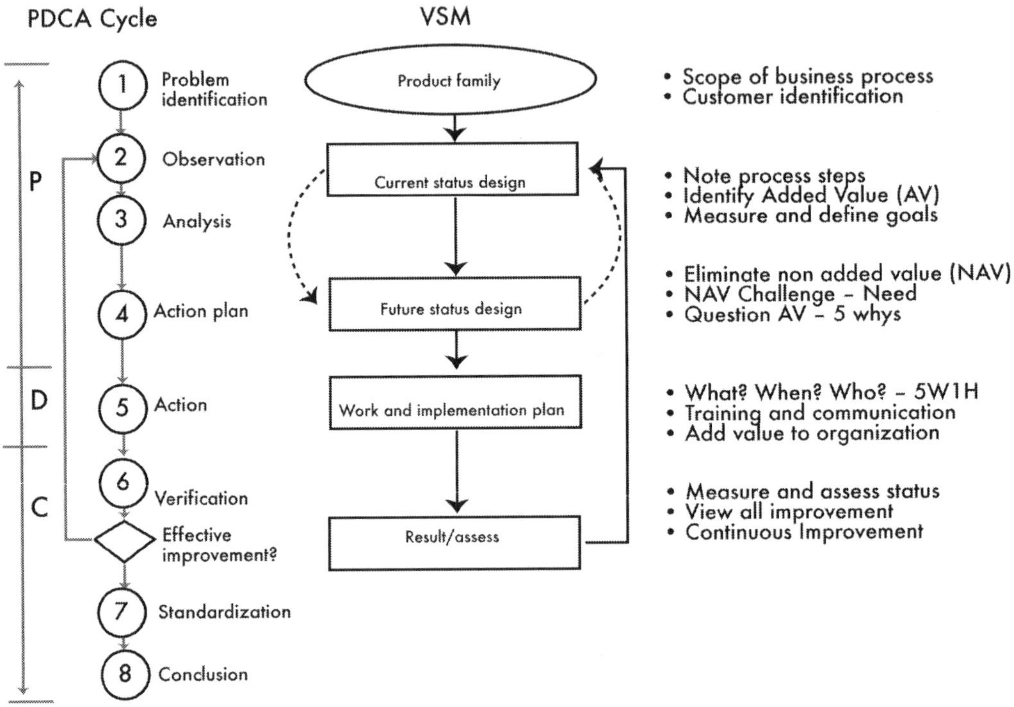

Figure 14 Value Stream Mapping and PDCA (adapted from ROTHER; SHOOK, 1998)

2.4.3 Supply management

The flow value is any action with or without added value required for processing each product, which is important to the production flow of raw materials to the consumer, and the design of the flow until its release. The total flow through the entire Value Stream towards the consumer and starts and the last layer and runs through the supplier manufacturing process(Figure 15).

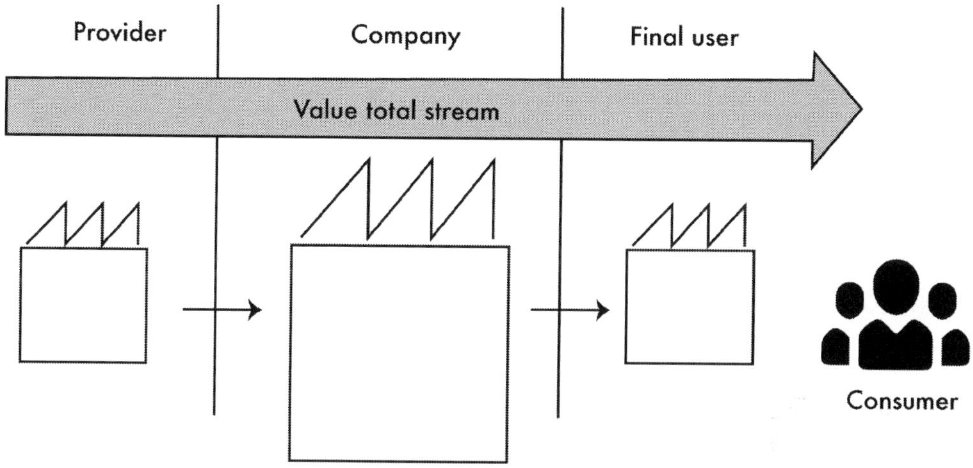

Figure 15 Illustration of the Full Flow Value (adapted from ROTHER; SHOOK, 1998)

There is already awareness from the manufacturing companies that the management of the supply chain has strong strategic implications and, through it, you can increase competitiveness, but there are concerns with the appreciation of the flow of products and/ or services and information flow that in the last two decades will go through a gradual change to apply the Value Stream mapping tool.

The Value Stream or supply chain is a network of services, materials and information flows that connect the customer relationship processes, order fulfillments and relationships with the company's suppliers and with their suppliers and customers in turn (KRAJEWSKI et al., 2009).

Logistics is the strategic management process of the purchase, transport and storage of raw materials, parts and finished products (in addition to the related information flows) by the organization and its marketing channels, such that the current profitability and future are maximized through the delivery of orders with the lowest associated cost.

According to Ballou (2001), logistics is the process of planning, implementation and control of efficient and cost-effective flow of raw materials in-process inventory, finished goods and related information from point of origin to point of consumption, in order to meet customer requirements.

Collaborative management is to reduce or eliminate the safety stocks that exist between organizations in a chain by sharing information on demand and current inventory levels. The focus of managing the supply chain is cooperation, trust and recognition, "…The whole can be greater than the sum of its parts", managing relationships, achieving a more profitable outcome for all parties in the chain (CHRISTOPHER, 2007).

According to the author, the supply chain management is the management of the amount relations (from the beginning) and downstream (to the end) with customers and suppliers to deliver more customer value at a lower cost to the whole supply chain.. It is still a network of connected and interdependent organizations (Figure 16), working together in mutual cooperation arrangements, to control, manage and improve the flow of raw materials and information from suppliers to end customers.

Authors such as Taylor (2005) and Correa (2010) present illustrations that represent the supply chain, in a systemic way, showing this interconnectivity. An example is shown in Figure 17, the main terms and the relationship of interconnections between operations of the supply chain.

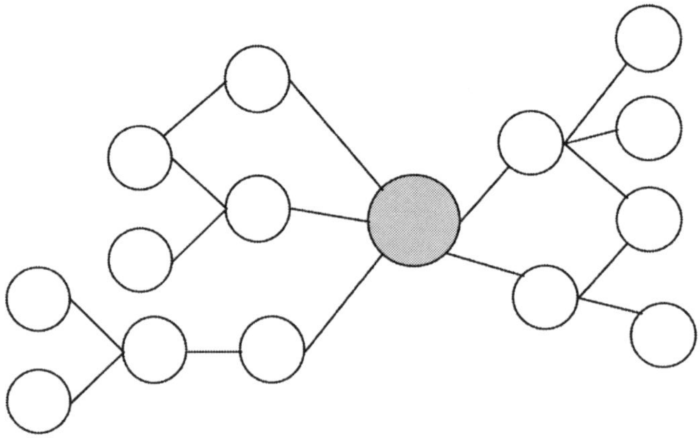

Figure 16 The Supply Network (CHRISTOPHER, 2007)

The enterprise (focal company) rests in the center of the supply network and companies B and C (focal level) are related companies or even competitors. To the right of the graphic we have layers of customers or even customer calls; to the left we can find layers of suppliers.. When the chain is being implemented the supply network is on the left side of the company A, the amount layers of suppliers go to the right of Company A, and downstream, we find the tiers of customers. The Illustration of products/services flow goes from left to right and the information flow goes in the opposite direction, the pointed lines are always from right lo left which shows the supply stimuli. (CARVALHO; PALLADINI, 2005).

The supply management describes the effective use of the purchasing market to meet demands in the production of the desired products, by reference to costs, advantages of quality and innovation. Under the business perspective, is the sum of all activities for the development of the supply of materials, facilities, services and goods. From the perspective of logistics, all activities that cover the demand of materials. Through supply management acquire the right parts of certain suppliers, the global market opportunities to take advantage, defines the responsibilities and interfaces with external suppliers (CHOPRA; MEINDL, 2003).

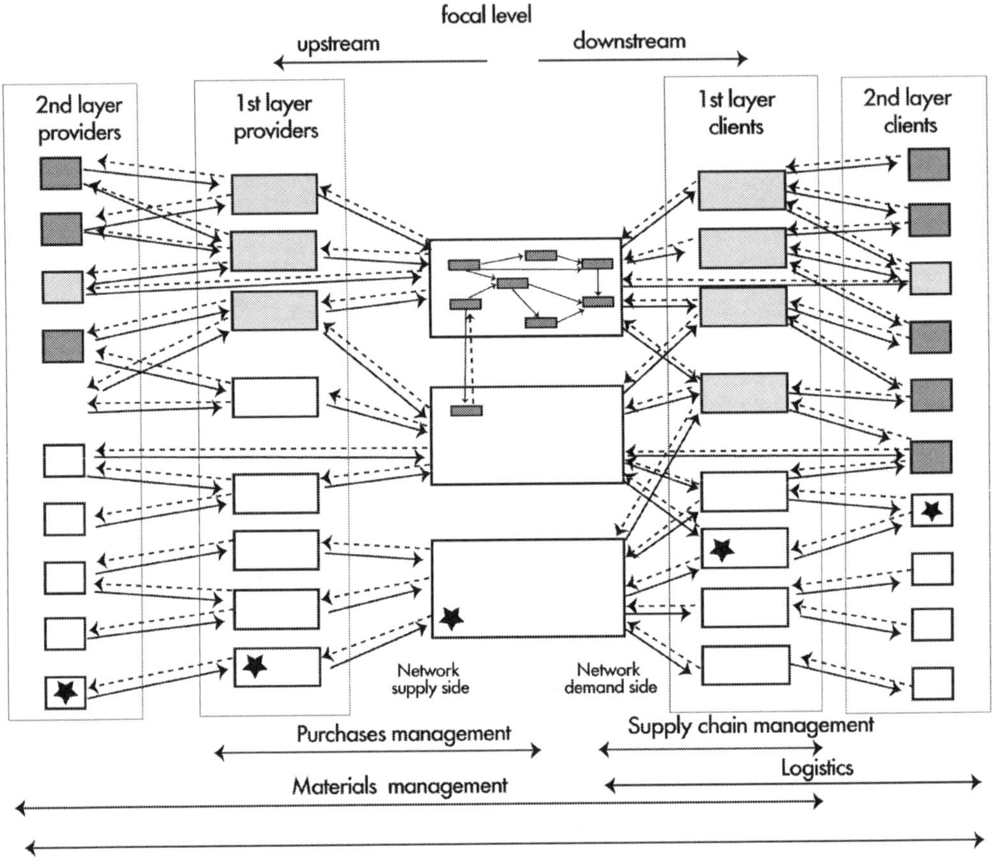

Figure 17 Representation of the supply chain network with interconnections and activities. (Adapted from Slack et al., 2002 e 2009)

The purchasing department operates by covering areas and functions (development, quality, availability) together to ensure the best combination of function, quality, delivery and cost. Purchasing decisions are made solely by criteria and objectives. The goal is to consider the total benefit (total cost) borne by all participants, that means among other things to also considered transaction costs and variable costs.

The global procurement network ensures an international market supply. The overall market cost level considering the total costs, is the greatness of determining guidance for all sites involved with purchasing and contracting services. The supply management in Figure 18 is characterized by a hiring differentiated services by type of demand, which is guided by the needs of consumers of business (manufacturing, assembly etc.).

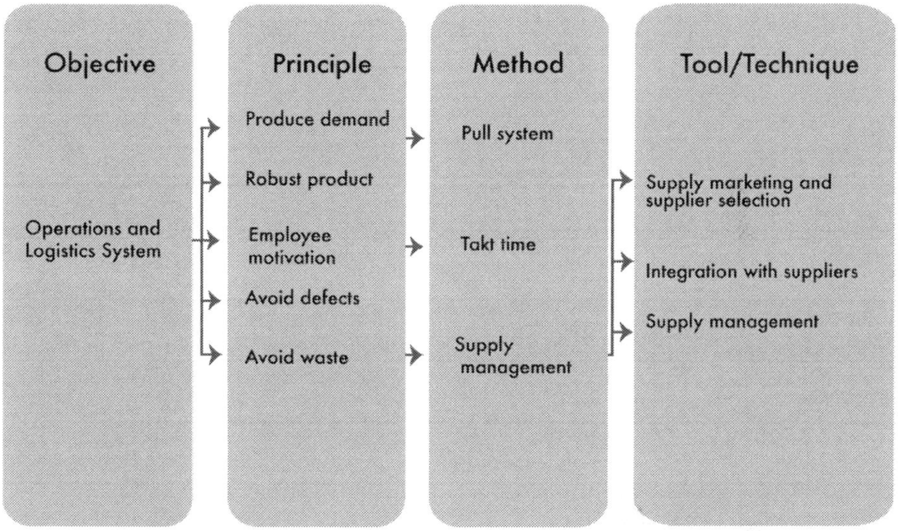

Figure 18 Tools for supply management in manufacturinga

All levers for contracting services are used, that is, commercial approaches guided by competition, as well as technical levers, related to the integration of suppliers and process improvements. Optimization of costs and benefits to products and services are performed, covering areas and functions. One way to achieve good performance in the supply management area is: select suppliers always looking for a way to communicate and integrate with its suppliers and assist management (LIKER; MEIER, 2007).

Marketing supplies happen when you have a development of the supply market, describing a structured process to identify the best supplier for a given part or material.

The selection of suppliers is understood as a systematic and reproducible selection of existing or new suppliers; this occurs within the context of pre-defined goals, that is, the strategy for material groups and strategy providers in their respective field materials. By assuming the total costs and optimizing the cost-benefit (material, logistics, quality and costs of internal processes) in regards to supplies we seek to minimize risks and supply purchases (SIMCHI-LEVI et al., 2010).

The integration of suppliers describes the business processes connection to supplier processes. Interfaces, responsibilities and procedures between the company and the supplier are defined and standardized by the Supplier Integration. Where necessary and possible in part the activities are transferred to the supplier, such as VMI - Vendor Managed Inventory. Thus, the know-how of the supplier is incorporated into the continuous improvement of the supply relationship.

When it comes to the integration of suppliers reconciling supplier processes with the requirements of the company are determined by the concepts of supply that merge together as: conditions for supply, warehouse and concept of ownership, control method and flow supplier of information is improved finally to use the company's resources optimization with the help of a feature that is used among other things as a tool to reach the integration in the flow of information is the EDI - Electronic Data Interchange (NOVAES, 2007).

The vendor management describes the guidelines for the systematic management of relations with suppliers. Important elements in these processes are the evaluation, classification and the development of suppliers, given that supplier management defines the relationships of the companies with their suppliers based on solid grounds, in which the individual strategies can be set in the context of the material group's strategy.

Consistent evaluation of all suppliers is available and is communicated to all the parties involved. Supply agreements are recorded in writing and are available. All relevant tasks (competence/responsibility) are transferred to suppliers.

As to external suppliers classification is feasible to develop suppliers / service providers with service providers through documents recorded in writing and available agreements relating to service levels. Suppliers are always audited according to plan which is a process already included in the procedures of companies that for example, have implemented ISO 9001, becoming more transparent business.

2.5 Robust products

The development of robust products requires the guarantee of efficient phase-in and phase-out processes with labor and products that meet the manufacturing requirements. This allows starting production within the deadline and with ensured quality. Some external factors disturb this process (such as changes after the application of the project) should be avoided or at least must be worked through a risk management and scheduling and takes into account the decision-making.

This also applies to the start of production due to the changes and shifts as well as the phase-out management and conversion to the production of spare parts. The purpose of this method is to ensure the quality, deadlines and control costs in a simple way to quickly place new products in the market and keep product changes at minimal cost. Remembering that this is a prerequisite for further growth of international competitiveness.

A good performance is attainable when it complies with the procedures to launch a product or even to modify it, so it is suggested to follow the PDP (product development process) consistently, which means in particular: keep a product project consistent, finalize the design of the product on time, complete the tests and clearances of parts, complete the logistics planning and quality assurance.

However, to ensure a degree of product maturity and processes at the beginning of the production phase there is a requirement for the selection and qualification of suitable employees; standardization of workplaces in the planning phase; integration and qualification of suppliers of equipment, components and parts, and the use of efficient technical processes as described in Figure 19.

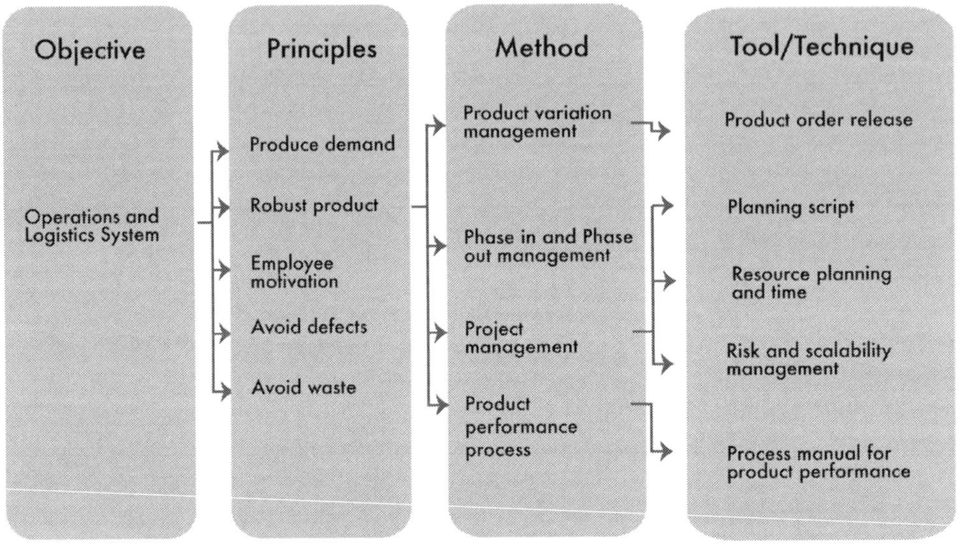

Figure 19 Tools and Methods of Robust Product Manufacturing

2.5.1 Management Variants and Modifications

The management variants and modifications describe methods for safe introduction of changes in the product; also guides the standard systematic variants control (MORGAN; LIKER, 2008) levering the domain complexity of the types of products and control variations and modifications through a consistent and transparent change process. These changes are systematically monitored for deadlines, quality and costs. There is a systematic process for regular control of all current variants, for example: EBIT (meaning gains before interest and taxes) and Quantities.

On the other hand the release of the production order, which is the initial release of the product to mass production ensures that all prerequisites for an ordered series production have been met for a product following the procedures of each company, for example: instructions, work plans, drawings, test plans, in addition to validation of parts, cost accounting and part availability (MORGAN; LIKER, 2008).

2.5.2 Phase-in and phase-out management

The management of phase-in and phase-out is a process for design of new variants, and variants of existing sets of changes. The control of the entry of new models (phase-in) and the output of older models (phase-out) without negative effects on supplies of skill ensures the planned start of the factory series avoids leftover products, raw materials and semi-finished equipment, sets out a plan of activities and deadlines to ensure the supply of spare parts.

Therefore, employees in charge of marketing, sales, development, product planning, manufacturing planning, materials management, materials management planning (via ERP - Enterprise Resource Planning), and quality assurance need to know and master the process and operate according to these procedures to improve the life cycle of products including recycling and disposal phases (MORGAN; LIKER, 2008).

2.5.3 Introduction to project management

Project Management is understood to be all regulations and methods to perform organizational tasks that are required for the development of a successful design. Corrêa and Corrêa (2007), point out that project management should include activities for goal setting, organization, planning, execution, control and monitoring of all tasks and resources needed to achieve the objectives defined for a project.

A project is an undertaking in which during a certain period of time a defined goal should be reached and that is characterized as essentially a single enterprise. The tools of project management method are described and used in the guidelines: Guideline for planning, resource planning and time and risk and escalation management. It is often applicable in development, manufacturing and quality management.

Resource and time planning mentor manage the resources available for the required work packages in order to achieve the objectives in an efficient manner and defined milestones. Eventually objectives, milestones, work packages or resources need to be adjusted to achieve optimal overall situation therefore becomes important to maintain the project goals and deadlines, it avoids multiple capacity planning, control is the logical connection between work packages, communicate responsibilities and follows up the progress of work during the project to avoid deviations on the cost, time and quality.

Risk management and scheduling are tools used to control hazards and preventive measures for the mere purpose of defending a project against unwanted effects. This management is used to identify and evaluate risks, and execute actions; action rather than reaction, and as swiftly as possible. This tool is used at regular intervals by the project team, under the moderation of their project manager.

Not all problems, difficulties and additional expenses that arise during a project planning processes can be predicted and prevented. Risk management and scheduling serve to assist in meeting deadlines and volume requirements, in compliance with the maintenance of project planning costs, investment costs and meeting quality objectives defined in the project proposal.

All activities are developed and controlled by a project manager who must be trained in risk management and scheduling, and must apply this knowledge continuously in the corresponding projects, along with the full team. Levels of escalation are defined as the risk classes resulting in a standardized process scheduling. Risks are classified for each project in a list of risks and procedures that are observed through risk class and measures that are accompanied consequently to the project's implementation.

2.5.4 Product development processes

This is the guideline for product development projects and provides the tools for project processing and proactively ensures product quality. With this process the cost and quality for the activities, methods, and objectives for the phases of the product can be defined along with the decisions thought out within a specific time frame . This is a way to guide the project to best practices, transfer knowledge in a uniform way for the company's employees (MORGAN; LIKER, 2008).

2.6 Challenge and encourage employees: people management.

Companies need their employees to work with thoughts and entrepreneurial attitudes, motivated to work with a high degree of responsibility but this depends on the company culture and work environment (Figure 20). Each employee in the workplace participates with a clearly defined responsibility, to achieve the company's goals. Task, responsibility and authority regulations are clearly defined according to the positions and salaries but it depends on the managers who must work with the staff to ensure that all employees are adequately trained for their tasks. Challenging and encouraging employees begins with the selection, monitoring and support for the continued development of employees. It's great when employees and managers are in constant dialogue and deal fairly between them (LIKER; HOSEUS, 2009).

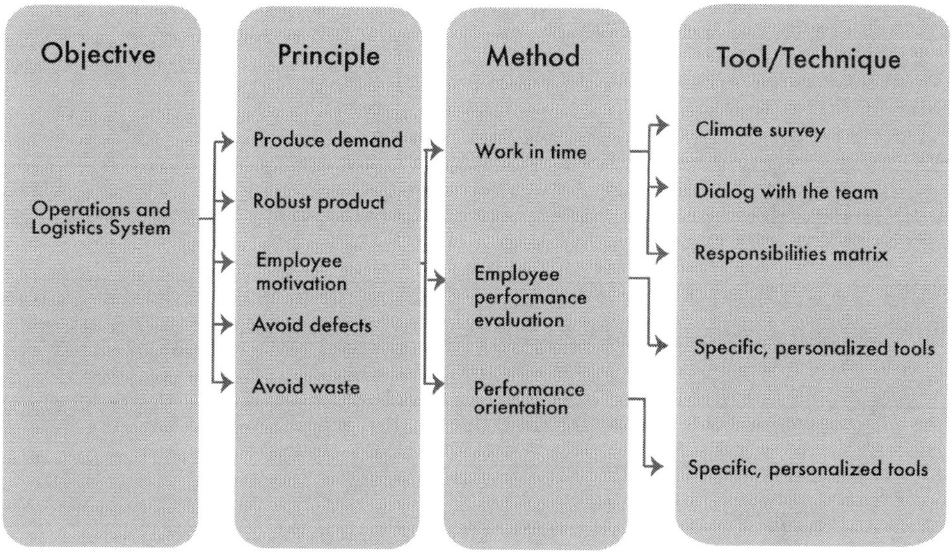

Figure 20 Tools and Methods for Motivating Employees.

Teamwork is the basis for cooperation and is expected to readiness for the work of all our employees. Oriented incentive systems for performance support to achieve goals are referred to the methods that human resources specialists utilize and select custom tools for each enterprise.

2.6.1 Teamwork

Teamwork is a form of organization where people work together to perform optimally complex tasks, in order to achieve the company's goals together. In the enterprise, teamwork assists for the concentration of each participant on common goals and reinforces the sense of responsibility of all. It also helps in the further development of all employees and increases a common responsibility; teamwork helps to ensure increased identification of employees towards their company and configures the company's success in this task in order to increase employee satisfaction, involving every worker in an optimization process.

Greater work efficiency is achieved through engaged employees and employees that co-operate with each other. Complex tasks are better mastered and coordinated when using the expertise of individual team members. Therefore one must inform and communicate in a standardized way, the whole group should discuss the problems and solutions whenever possible (CHAVES, 2006).

The working environment survey is a tool used to assess leadership behavior of a manager and to ensure the development of shared solutions to improve the working environment and collaboration within a unit of the organization. The dialogue between management and subordinates employees is moderated by a trained person, with the help of a systematic and standardized process. The climate survey should provide assistance and self-assistance. Before or during each shift employees talk and dialogue with the team in a brief exchange of information. Information exchange is needed for daily work, distribution of jobs, achievement of objectives, serious errors, current quality situation, new variants, scrap, rework, number of employees (workers on vacation, sick). As a suggestion there can be a dialogue with the team once a day for five minutes.

In the matrix of responsibilities, recurring tasks and activities are clearly documented, distributed to groups of people responsible, the type of responsibility is defined and, if required, the task execution frequency is recorded and documented. Who does what? How and When? Responsibilities, implementing tasks is visualized control. The responsibility matrix is created for all activities and recurring tasks. Documentation is available for each group and is displayed publicly. Each employee knows their tasks and follows the designation by the array of responsibilities.

2.6.2 Qualification and development for employees

Employee development contain every measure used for personal and professional development and measures that encourage the undertake, in the best possible way, of your current and future tasks. Qualification is an essential element in the development of employees, as it is understood by qualification, the process of achieving and developing capacities to fulfill a certain task or requirement. This is a way to ensure competitiveness, especially with the improvement of efficiency, quality process and safety, keeping the staff innovation capacity and ensuring their employability.

The enterprise has the advantage of ensuring the availability of employees, managers and qualified specialists, extending the maintenance of knowledge considered as enterprise value, and used to encourage people to develop and increase cooperation within the company. Thus, the enterprise reduces the technical and economic risks, develops employees and creates a self-fulfilling environment and job satisfaction.

2.6.3 Performance guidance

The company wants to increase its level or be present as a performance leader in products, processes, quality and costs, but it depends on a culture of excellence. Performance orientation identifies the ambition of the company and its employees in systematically focus to achieve the company's goals and achieve great results working. This method is also personalized and seeks to ensure the maximum contribution of each individual employee to achieve the company's goals and to increase work efficiency and avoid waste. Activities and tasks of content can be defined properly and be known by employees. Remuneration systems geared to performance and / or success is a project already tried and can be implemented according to the needs and maturity of the company.

2.7 Avoid waste: day to day routine management

According to Quality assurance procedures various methods and tools can be used to achieve a quality product that meets customer expectations, creates trust and builds relationships with customers and suppliers through standardized and lean processes for greater customer satisfaction, avoid non-compliance and achieve business objectives, such as reducing errors, waste, rework and warranty costs. In this document, a few methods used in studied companies are cited, illustrated in Figure 21 and described below.

2.7.1 Defect prevention

Defect prevention means identifying defects sources in the development of products and processes, prior to serial production to prevent their occurrence by implementing appropriate measures.

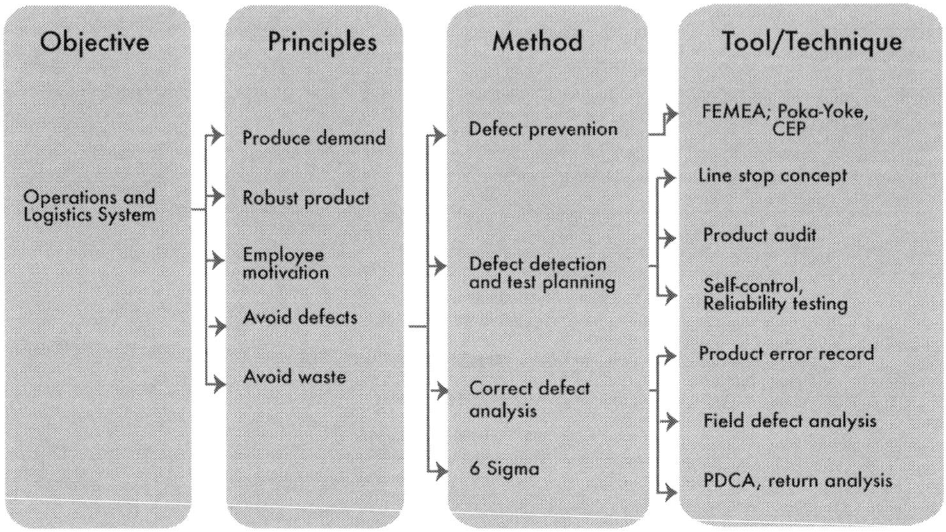

Figure 21 Defect prevention tools and methods

The defect prevention can also be activated by the analysis of defects found, lessons learned. The later a defect is detected and corrected, the more expensive it gets to the enterprise, given that every defect costs time and money. Preventing defects in the product's development stage may involve suppliers, especially in the development phase, the employees are aware of past defects hence it is feasible that defects become systematically documented and treated preventively. The tools of QCC (Quality Control Circle) have been a good alternative, but you cannot give up the knowledge and application of FMEA, Poka yoke and ZIP explained follows (GHINATO, 1995).

Failure Mode and Effects Analysis – FMEA is a formalized analytical method, to identify potential weaknesses, recognize their meaning and prevent in advance. In practice, there are the FMEA system (for new products, processes and interfaces); FMEA Design (for new components and parts); FMEA Process (for new manufacturing and assembly processes). The purpose of applying the FMEA is to increase the functional safety, improve products, reduce defects costs, scrap and rework, analyze the defect evaluation purposes, prevent repetition of defects and duplication of work, optimize system/ construction/processes create a knowledge base and document the experiences. FMEA teams are made up of representatives from all areas of competence after completion of FMEA from an RPN value (limit set by the factory) measures are planned in which they are implemented consistently within the defined deadlines. This method is used most frequently in the introduction of new technologies, materials or processes and systems, relevant parts or processes for safety.

Poka yoke is a method for preventing and/or prevent failures (errors/defects) caused by people accidentally. The translation of the Japanese corresponds to "Poka" = involuntary error "Yoke" = avoid), the integration of poka yoke in processes, products and operating means exclude the possibility of faulty operation or assembly. Poka yoke is a technique that facilitates the process of work and increases safety in the workplace and prevents damage to machinery, in some cases applies in assembly processes in which there is likelihood of confusion between components or potential misuse.

The statistical process control (SPC) is a preventive strategy of quality assurance for control and/or orientation of critical manufacturing processes, based on statistical methods. Its prerequisites process stability, given that the behavior of the process of assessing machinery capacity (CP) must be known, as well as the behavior of the evaluation process capabilities (CPK). The SPC has helped in continuous improvement capabilities of machines and processes, reducing costs for monitoring quality and has contributed to the quality certification ISO 9000. All documented critical processes and important (e.g., process map) are identified by CP and CPK and secured with zip code. In CP (Cmk> 1.67) in CPK (Cpk> 1.33), those responsible for processes are defined, measuring instruments are checked at regular intervals.

2.7.2 Defect detection/Test planning

Defect detection means that a deviation is detected in relation to the defined specification. The detection of defects is an essential prerequisite for the correction of defects (GHINATO, 1995). Each employee must inform and identify (mark) defects discovered in their work because customers must receive a product unblemished. Visual management is used to make the defects visible quickly. There must be a specification and plan for every process, as well as education plans and the plans to assess general conditions for testing that must be defined and maintained (e.g. Test means and capacity, lighting and noise level).

Another important concept is the production line stop, which is a stop in accordance with a defined procedure, when a problem occurs (e.g., performance, quality, missing parts etc.) and measures are implemented to prevent a recurrence of the problem. Each employee knows how to cause a line stop and what should be done in this scenario. Responsibilities for production line stop and the permanent solution of the problem are clearly defined. Medium and long term measures for permanent disposal problems are documented. The product Audit management should have corresponding authority (e.g., product ban, release) and any defects found are documented and corrected via PDCA or similar problems solving mechanisms.

Test concept means testing as actual nominal comparison regarding specifications, for example, the areas of development and quality management. To identify errors, avoid costs of non-compliance, product liability risks and prevent damage to the image of the company (e.g. Product testing). Only flawless products and manufactured according to customer requirements and standards leave the factory. The test content is described and documented in detail; these contents are object according to relevant departments (e.g., development, manufacturing, quality management).

Only authorized personnel can perform rework and grant special releases, and if deviations occur, these should be clearly documented. The test results are analyzed and routinely referred to a PDCA (including scheduling mechanism). Testing facilities are maintained and calibrated regularly (monitoring tests means). The staff responsible for testing is qualified and routinely participates in training (JURAN, 2009).

The employee controls his work, auto controls, with reference to the correct execution of critical steps mounting, as described in standard worksheet. Only products free of defects are passed through the next stage in the work process. Increased awareness of quality by the employee, and greater involvement of employees in defect liability helps to correct the defects promptly.

Long-term product reliability tests accompany the manufacturing process and the samples are representative. Finished products may come from own production or third parties. The tests should, for example, simulate guarantee time or durability, and ensure product quality beyond the warranty period and life over the life of the product. Identify system failures, leading causes of faults and long-term quality trends and product optimization for implementing measures.

The testing concept implements customer requirements, legal requirements, and internal authorities for reliability testing. Content, means and methods of testing are defined and documented as concept testing. The evaluation of defects is performed according to standard company - weighted rate of defects in Audit Product. The priorities of the defect classifications (A, B, C) are clearly defined. Test results are recorded and evaluated in Pareto chart representations, at least once a month and informed to the factory management. The opinions of those responsible are brought in due time.

The Pareto analysis is a concept discovered by Italian economist Vilfredo Pareto, which found that about 80% of the world's wealth was in the hands of 20% of the population. This principle of 80/20 is widely used in the analysis of various everyday situations operations. In the 1960s, Juran proposed the Pareto analysis as a way to "separate the vital few elements" in an analysis. It is understood that sorting problems in descending order starting with the ones that produce greater effects and attack them holds a greater solution capacity and maximizes the results (BRAZ, 2008).

2.7.3 Analysis/Correction of Defects

Correction of defects means that specific deviations are corrected in the shortest possible time. A rapid and lasting correction of defects requires a structured process, clearly defined responsibilities and consequent monitoring and measuring the success of implemented measures (Shingo, 1996a). The later a defect is identified, the more expensive it will be to fix. Correction of defects is systematically executed by PDCA tool by involving relevant areas. The sustainability of the measures implemented is ensured by plans of corrective measures, status monitoring and measured variables and responsibilities for the correction of defects are defined and communicated.

Production error log is a record full purpose of defects in the production process of multinational companies such as BSH, Bosch, Siemens, Eaton and many others. The FPY indicator - First Pass Yield - is based on the number of products and/or assemblies that have suffered rework due to defects thus providing information on the share of products that travel the production process without correction or rework. This type of record is used to optimize workflow, identify deviations and prevent the costs caused by defects and product liability risks, and a way to control the identified deviations. It is noteworthy that the processing occurs on a routine and standardized by the use of appropriate techniques of problem solving.

Defect analysis in the field, meaning problems occurred with a client product (end user) during their entire life cycle is important. It contributes to identify weaknesses that were not detected before shipment by the manufacturing company thus, preventing the costs caused by deficiencies, warranty costs, and damage to corporate image, identifying the potential improvements (Slack et al., 2008). The main defects points are systematically filtered and detailed precisely through analysis of individual documents relating to repairs. The results are documented and can be submitted monthly through graphics. Prompt notifications of incoming defects are received in customer service so they can give out a detailed response and the time since any defect found is documented and corrected via PDCA.

The problems are described and analyzed exactly as the process of PDCA - Plan, Do, Check and Action. With containment measures, the effects of defects are identified, and/or eliminated. In addition this ensures the elimination of long-term causes of defects. PDCA's ensure that, in spite of a drift, the quality objectives are achieved through measures and procedures.

Engage the next higher level in case of possible problems or deviations that cannot be resolved at this level and the information products and processes are communicated to responsible areas to determine any changes in the manufacturing process or product. Given the practicality and results in companies the PDCA troubleshooting method began in the areas of quality, but it is used in the areas of processes and services of companies of any segment.

The return of analysis of parts, components or products that have failed in the factory: the customer is a participant that allows companies to contribute to the improvement of quality because they identify current defects in parts, to eventually be able to react even before shipment of the products, shipping lock or to use existing stocks, rework and return.

It will be the proper disposal of defect causes (supplier, assembly, logistics etc.) to fair assessment of supplier and determine correction/improvement actions. Determine the supplier's share of responsibility by a defect as a basis for the calculation of the cost of warranty claims. The defect identification mechanisms in the field should be used as a basis for improvements in new developments.

Defective parts that are the supplier's fault are returned to it, along with the 8-D report and processing improvements measures that are controlled; defective parts per supplier's fault (internal only) are inserted into the ERP or similar production system. All parts that fail within the plant are evaluated at appropriate intervals, these results are immediately communicated to all involved departments and corresponding actions (inventory control, shipping lock finished products) are implemented immediately.

2.7.4 6 Sigma –boxplot chart

Six Sigma is a methodical process for systematic enhancement or re-designing processes to increase quality and efficiency. Companies use Six Sigma to ensure sustainable process improvements, compliance with business requirements/economic accounts, seek increased customer satisfaction and loyalty/long-term customer retention with an incentive-oriented thoughts variants. Six Sigma is anchored in the organization, so all six sigma projects are documented by continually monitored monthly reports and are informed to the factory management. Theories and 6 Sigma practices are comprehensive, so it is noteworthy that this work explores the boxplot tool Minitab software.

Among the basic quality tools that can be used for preliminary analysis of the process and seek to start the analysis of the obvious causes of variation and potential, according to the approach 6 Sigma is the Box plot. This tool is available in Minitab software and other computer programs. The box plot is a graphical representation of the distribution of data variables that presents symmetry and variability data. The boxplot summarizes the best that can be dispersed with the measures involving ordered data. The values are arranged in ascending order with 100% of the time, this percentile in the order 25, 50 and 75 are called quartile and represented by Q1, Q2 and Q3. The boxplot consists of a box (a rectangle), as shown in figure 22, with the upper level set by the third quartile (Q3 = 75%) and the lowest level defined by the first quartile (Q1 = 25%).

The median is represented by a line within the box (Q2 = 50%) and the maximum and minimum point are connected to the housing by straight segments. The outlier can also be viewed, a discrepancy value and away from all the data, so it is an extreme value, which lies well outside the general pattern of almost all other data. Most common forms of sample distributions are symmetric distribution (see Figure 22); asymmetric distribution to the left; asymmetric distribution right and uniform distribution (BRAZ, 2008).

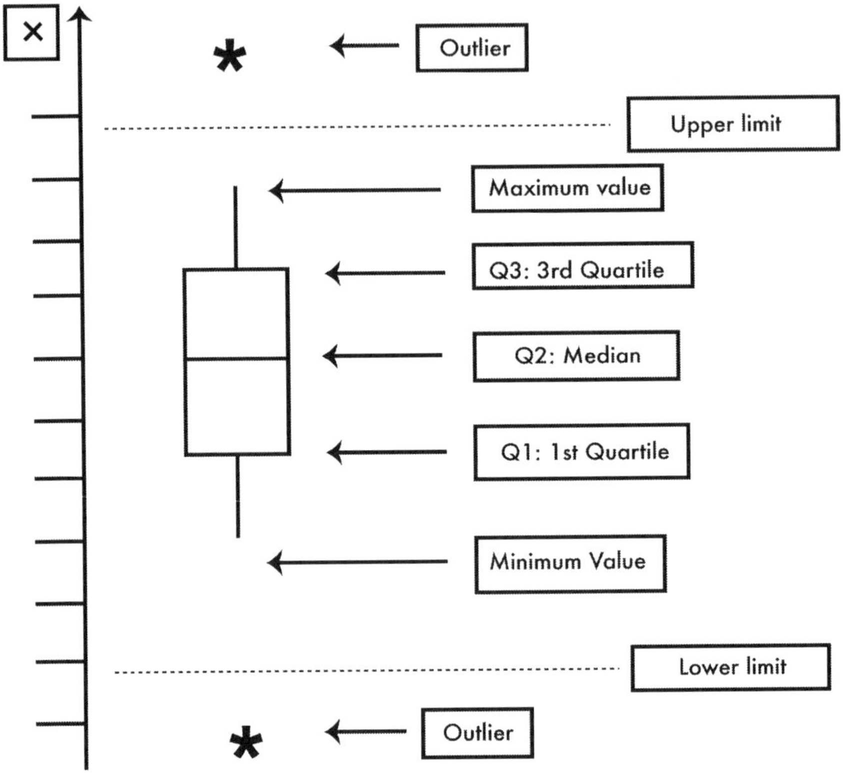

Figure 22 Boxplot for a single set of data (BRAZ, 2008)

For Triola (2008) box plots are particularly useful for comparison of data sets, especially when they are drawn to the same scale. Once viewing simultaneously several important aspects of the data, such as central tendency or position, dispersion or variability, departure from symmetry and identification of very distant observations of most of the data (such comments are often called discrepant or outlier) one can compare all the same data in different times and see how much each boxplot is better or worse under the aspects mentioned.

2.8 Avoid waste: operations and management services

The operations function is the organization's set of activities that produce goods and services. Each organization has an operative function because every organization produces a mix of product and services. Processes are the components of operations. The lack of management operations, processes and services can improve or break a business and according to Slack et al. (2008) when they are managed well these management operations can contribute to the enterprise's strategic impact in four ways: cost, revenue, investment and capacity. The function operations shall maintain and control costs, provide products, service and quality, manage capacity and generate revenue, always with the lowest possible investment in a sustainable and competitive context.

Slack et al. (2009) suggests a model in Figure 23 in which the relationship of operations management with operations strategy is shown. The operations strategy is guided by the objectives of operations strategy, these come from the breakdown of targets and company guidelines.

The management operations and services should cooperate and be aligned with the projects, with work related and focused to achieve important results, which are usually unique and has a pre-set time, as they seek to reduce the costs of products and services, increase revenue with quality, service and innovation to achieve the customer's perspective, also seeks returns on investments and build a responsive competence based on knowledge and innovative skills.

Figura 23 Gestão de operações (SLACK et al., 2009)

Among the many responsibilities of operations management services and the management of day-to-day routine seeks to guarantee the quality and support continuous improvement. It also manages the ability of equipment to perform better planning and control of funds from operations in order to transform and meet the demand for products and/or services and to the best attainable performance with the volume size, variety, variation demand and visibility.

Within the current concept of sustainable and competitive business management, it is necessary to have a continuing search for operations that add value to the customer and avoid waste. In this context is favorable to apply the methods and Lean tools to maximize the value added by each of the activities of a company through the elimination of unnecessary features and excessive delays.

The mindset described is to avoid waste, as the customer refuses to pay the value for minimal continuous improvement which means ensuring the best use of resources through the results orientation and sustainable processes, (Figure 24). The Creation of procedures focused on adding value already in the development and planning, and eliminating waste, on their own, are not only necessities in production but in all areas along the Value Stream.

The aim is to increase total productivity, through increased productivity of direct employees, increasing the OEE, reducing fixed costs etc. The procedure is to standardized continuous improvement processes, resources, equipment and operating means, so deviations are visible and eliminated through the waste. .

Attain operational sustainability. After each improvement defines a default, their performance level is maintained and measured. There are eight types of waste: transport, waiting times, drive, over-production, rework/defects, unnecessary processing, inventory and, finally, the lack of use of people's intellect and creativity.

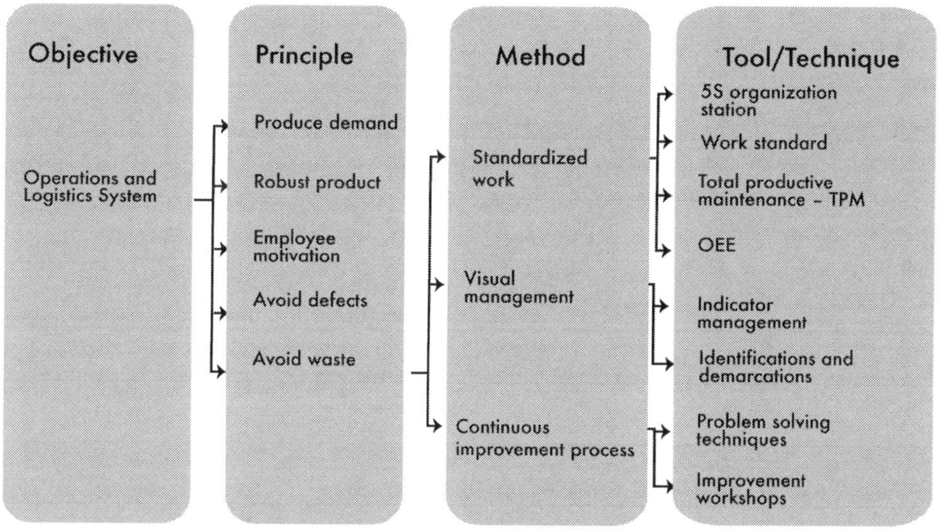

Figure 24 Waste avoiding tools and methods

2.8.1 Standardized work and introduction of the OEE concept

Standardized work defines explicitly the running and current process in which a worker performs. The work must be performed always in the same way, regardless of person, place or time and each individual pattern represents a particular time, in the best and safest way to perform a task, noting that the standardized work becomes transparent through the management visual.

The advantage of standardized work is to comply with standards to ensure sustained efficiency and transparency of work processes, by making existing deviations visible. It increases continuously improve security and stability of processes and promotes the skills of employees. This method needs the involvement of employees and initiatives to improvements (CALADO, 2006).

The organization of the work environment (Japanese: 5S) creates a local concise work, ergonomic, continuously clean and safe and increases the motivation of employees and the hence the efficiency of it. Transparency is achieved, search times and long paths are reduced, and the use of space, availability of equipment and staff motivation are improved, not to mention that minimizes the risk of poor quality. Order and cleanliness are visible in all areas; the working standards and their locations are clearly identified and arranged based on their effect of creating values. Audit system responsibilities are defined by cleaning checklists and plans.

The creation of worksheets or documents uniformly standardized that define production processes are the basis for simultaneously identifying deviations and improving the workflow, they are means of visual control. The standardized worksheets obey the customer takt (takt time) and are based on a balancing operations chart, also called load graph of labor. The reason to have this procedure is to document standard workflows, ensure uniform implementation of the work and quality standards, continuously improve processes, based on standardized workflows, accelerate the education and training of employees and introduce routine optimized, create basis for identifying deviations, view work contents that add value or not and also creates the basis for communication between employees and managers (CALADO et al., 2005).

Total Productive Maintenance (TPM) describes a holistic concept of maintenance oriented production, to increase the overall efficiency of equipment, Overall Equipment efficiency (OEE) and reduce maintenance costs through active measures of preventive maintenance, with the involvement of all participants. Only machines with reliable facilities are reliable to perform an efficient manufacturing streaming therefore, seek for continuous equipment efficiency, increased reliability and simplified maintenance as well as reduced maintenance costs because it is understood that equipment with a good condition can manufacture high quality products. Performance indicators are integrated in the operational processes, OEE losses and sources are recorded, monitored and viewed and the audit TPM measures are performed at regular intervals. The application occurs frequently in the areas of production, machinery, equipment and operational processes (CALADO; BIONDI, 2007).

Machines and humans are prone to failures that can lead to errors and defects, affecting systems, facilities, personnel, suppliers and customers. With the implementation of appropriate measures, failures can be reduced and forecasting defects can identify the root causes. Thus, conditions that ensure standardized, reliable and efficient processes are sought, in order to provide quality products that meet not only the customers' expectations, but also avoid interruption of the process due to equipment downtime. The defects generate instability in low EEO process and end up committing the time and cost.

It can therefore manage a production system through the OEE and know the incidence of problems and / or outbreaks of losses and wastage in the product flows, processes and information in real time. Having henceforth indicative of OEE in real time, you can then develop an action plan based on the use of focused methods to combat the origins of waste in production processes (CALADO et al., 2010).

In 15 years the OEE researched in the international database Emerald, from January 1994 to September 2009. 84 international articles were selected, in which 83% of these were published in 12 journals (magazines and newspapers), making their articles highly relevant.

As a result of the research a growing importance of OEE for academic and industry in the last two decades is shown (Graph 2). From 1994 to 2003, it reached an average of 2.8 articles per year and reviewing the latest statistics from 2004 to 2009, the last six years, it is possible to see three times as many articles published, i.e. 2 8 articles are averaged 9.3 international articles per year dealing with the OEE and other topics that are related.

OEE (Overall Equipment Effectiveness), treated as Global Equipment Efficiency, can be understood as a system or program, but it is an important metric used in different industries, and has become a global performance indicator for organizations to measure the productivity of industrial operations (HANSEN, 2006).

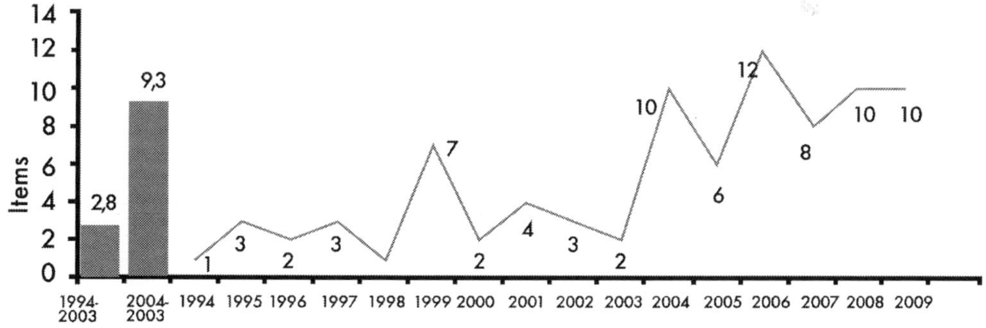

Graph 2 Articles on OEE – journals from 1994 to 2009

According to Dennis (2008), the OEE major losses related to manufacturing equipment are due to the following states: 1. without program, 2 out of the program, 3. shutdown, 4. improvements, 5 and 6 productive auxiliary as figure 25.

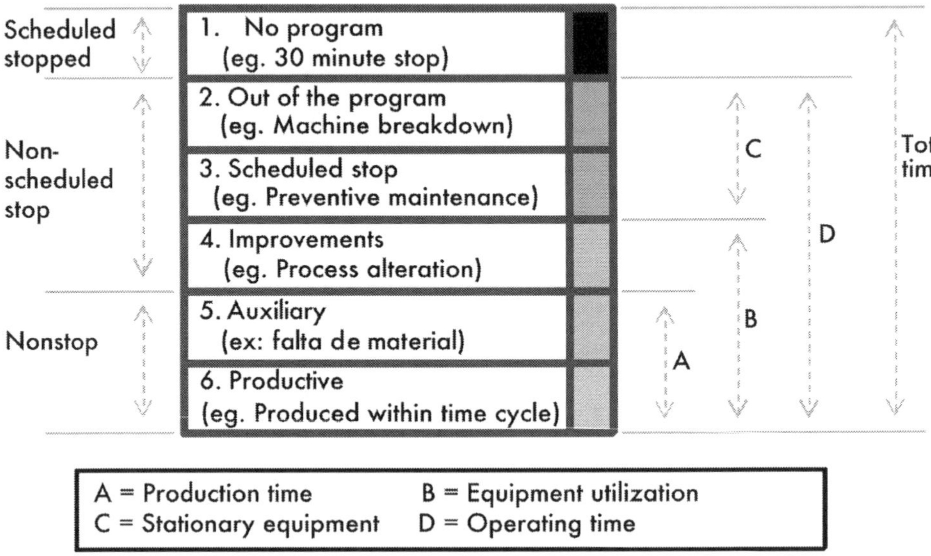

Figure 25 OEE status (adapted from NAKASHIMA, 1988)

No Program Status: means lack of programming, it is considered as a scheduled stop, that is, when the equipment to be used in production stops on a scheduled basis. Examples: stop on weekends, holidays, under tool changes and downtime of the company (union meetings).

Unscheduled Status: classified as an unscheduled stop, is when the equipment is unable to perform its planned function due to unplanned events. Examples: stop for maintenance, i.e., the machine is waiting for personal or parts, awaiting repairs or stop to exchange the material consumption.

Scheduled shutdown status: been mentioned as unforeseen failure, is when the equipment is not available to perform its intended function due to setbacks maintenance. Examples: This state includes activities such as production testing, preventive maintenance and organization.

Improvement Status: classified as an unscheduled stop, is when the equipment is in condition to perform its intended function, but is operated to administer experiments improvements. Example: this state includes process engineering and maintenance activities to improve and modify the equipment.

Auxiliary Status; considered non-stop, i.e. when the equipment is in a condition to perform its intended function, but it is not operated Examples: when there is no operator available (absenteeism, lunches and meetings), there is no material available or no tool support. Productive state, is also considered non-stop, i.e. when the equipment is performing its intended function, which can be classified in the following states: regular production (including the loading and unloading of parts), work for others, rework or improvements made along with production. The six states of EEO are fundamental exemplified in Figure 25.

Once the states of OEE are defined, the next step is to discuss four types of losses. The first type is the loss planned (e.g. standstill in the days of holiday), the second type is the operating loss that are still affecting the availability of equipment, and this occurs when the equipment breaks or there are delays that disrupt transformation process, as in the case of exchange of the tooling, the third type is the loss of speed, given that when you change the speed of the production for whatever reason this affects their performance and productivity directly by modifying the desired result, being able to cite examples of such loss: short breaks and abatement equipment speed.

The fourth type of loss with respect to the loss of quality because the defects and flaws in the product generate waste at the beginning of production and rework. According Nakajima (1988) that implemented the concept in hundreds of Japanese companies in the seventies and recent authors' state that the OEE is calculated by multiplying a rate of availability for the performance rate, quality rate, as shown in figure 26.

As improvements in the management system can be considered online data collection of the production process and availability of information in real time, generating key performance indicators, such as: MTTR (Mean Time to Repair); MTBF (Mean Time between Failure) and TEEP (Total Effective Equipment Productivity).

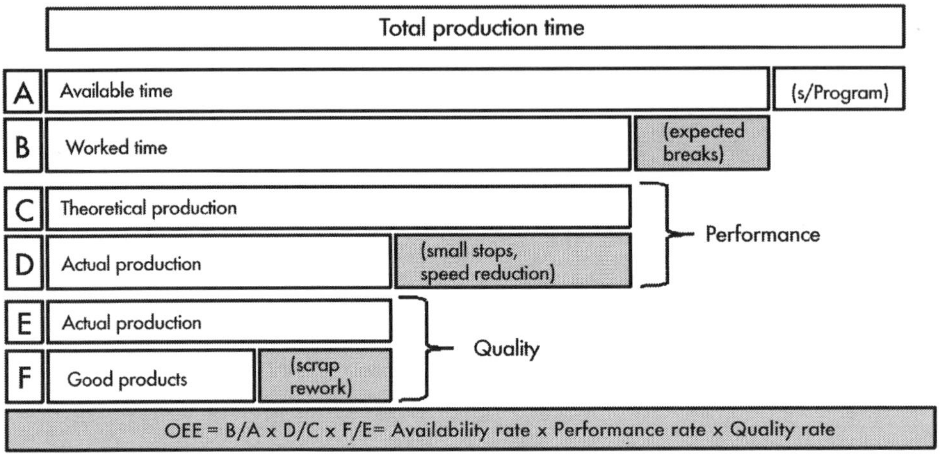

Figure 26 OEE Calculation scheme (adapted from NAKASHIMA, 1988)

2.8.2 Visual Management

Visual management means visualization processes, workflows and results for employees, visualization tools are coordinated to your target group and contents, as an example cited there were possible means of information: posters, information boards, books, movies etc. The goal is to make processes transparent and easy to understand content to achieve clearly identify patterns and deviations, accelerate the solution of the problem identification, support and motivate employees in fulfilling their tasks and achieve its objectives, inform employees about: history, current situation, business objectives and increase the identification of employees with the company, the desktop, your task and the product. The performance indicators of management tools (KPI Management) and identification and markings, are feasible and applicable throughout the company when it has increased awareness of the benefits.

System performance indicators, also known as performance indicators, fragment business objectives in areas, enabling employees and managers to identify how they can contribute to achieving the company's goals. Key performance indicators (KPI - Key Performance Indicator) are tools to measure the results achieved. The management indicators describe the selection, determination and visualization of key performance indicators, as well as the means to control this data. Management uses the indicators to help present and monitor goals, visualize trends, evaluate and optimize processes, identify problems quickly and encourage communication and information in the company. This action helps in case of deviations from targets where measures are introduced and its effect is monitored on the basis of performance indicators.

Identification and demarcation are efficient means to implement work procedures in production. These items classify, detect and signal objects and areas to identify patterns of deviations in a clear manner, increased transparency of work procedures, material flows and stocks. Giving quick and easy orientation in the workplace, with training for employees with their goal to understand the signs and marquees.

2.8.3 Continuous improvement process

The continuous improvement process (CIP) is used for the continuous and consequent enhancement, in small steps for all the company, involving all employees, figure 27. The improvements achieved are recorded as standard, to ensure its incorporation into the process and eventually your transfer to other application areas. As with kaizen what the desire effect is to increased productivity and quality, reducing waste, rapid implementation of improvement measures, improving workflows and re-designing jobs to achieve shorter lead times with staff motivation through the implementation of their ideas and their integration into the company. The continuous improvement process (CIP) enables further improvement between the jumps represented by innovations.

Troubleshooting techniques are standard systematic procedures used to identify the root causes of problems in order to eliminate them completely. Thus, we analyze the causes and solutions of resulting problems, reduced time between identifying and eliminating or correcting problems. The troubleshooting is performed at all levels, with proper tools for each situation.

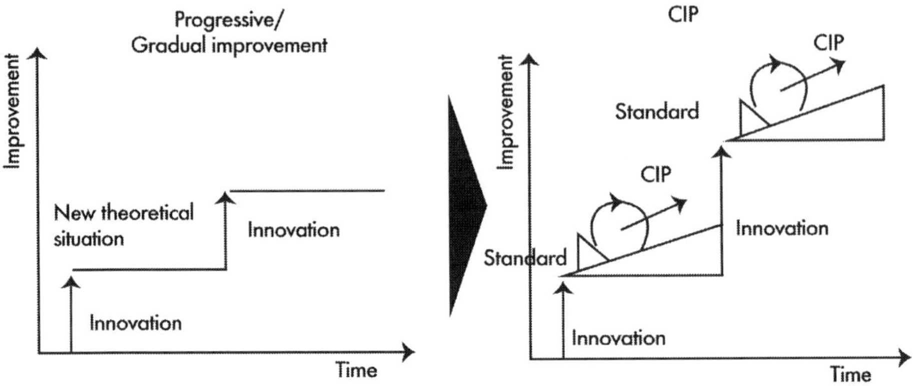

Figure 27 Continuous Improvement Process (adapted de CAMPOS, 2004b)

Apply the tools for continuous improvement of processes and workflows, with the assistance of workshops with the participation of the employees is an alternative to favor the employee's involvement in the continuous improvement process, generating an increased willingness to implement improvements and encouraging participation of employees. Workshops are run to plan and according to the need of continuous improvement, there is an opportunity for participants to present the results to managers and this facilitates implementation with on-site viewing and monitoring of results.

3. THEORETICAL BACKGROUND ON BENCHMARKING AND HOSHIN KANRI

3.1 Industrial benchmarking

Benchmarking can be defined as the search for best practices of the company leading to superior performance (CAMP, 1989). Launched by Xerox, benchmarking has been widely adopted since the 1990s by companies as an improvement initiative. According to Voss et al. (1997), Benchmarking is vital for the company to learn how to improve its performance and the organization, the improvement can be achieved through the identification of good practices and establishing performance goals. This model (Figure 28) seeks to increase understanding of the strengths and weaknesses relative to competitors and you can get real benefits as they focus on the organization's needs.

The learning orientation has been linked to benchmarking and supports transformation of the production system. Voss proposed a relationship between learning, benchmarking, understanding and performance, summarized in the research model (Figure 28).

Industrial Benchmarking was developed by the London Business School, in partnership with IBM, associated with the CBI (Confederation of British Industry) which maintains an international benchmarking program, accounting for a database with more than 1000 companies from 34 countries (HANSON; Voss, 1995).

The Euvaldo Lodi Institute of Santa Catarina - IEL / SC is accredited by the International Comparison Ltd. to implement the industrial benchmarking in Brazil and, with the support of FINEP, is accrediting a network of peer institutions through the Best Practices Program for Industrial Excellence - PMPEI, the IEL has application of partnership with the Information Technology Center - CTI in Campinas, São Paulo. Thus, it becomes possible to offer a unique tool to the domestic industry for evaluation and comparison of their practices and their results in relation to world leaders.

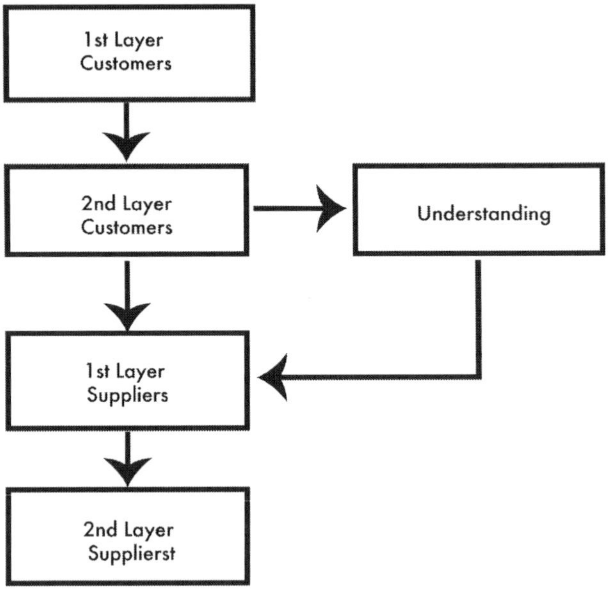

Figure 28 The benchmarking research model (VOSS et al., 1997)

This benefits enterprises with the possibility of identifying and highlighting the strengths and weaknesses of its management, subsidizing investment decisions and providing strategic information on the sector in which they operate. It is noteworthy that all information about the company that hires an application of industrial benchmarking is kept under strict confidentiality. The individual data will be disclosed only with the express authorization of the company. Participating in this program allows the companies to take another step in the search process for manufacturing excellence.

3.1.1 Benchmarking methodology and steps for the team

The goal is to help maximize the company when using the industrial benchmarking. The enterprises must meet the three key components that will be involved in the application of industrial Benchmarking:

The industrial benchmarking facilitator, who is the consultant accredited by IEL / SC to apply the industrial benchmarking.

Benchmarking team leader, the person responsible for coordinating the application of industrial benchmarking by the company.

Benchmarking team, which is the team made up of people of various levels and various sectors of the company to follow these steps:

Step 1 - Answer the questionnaire, receive the questionnaire and answer as many questions as possible, giving individual score from 1 to 5. Discuss with team members and management how is the score a reflection of their area and search for examples or evidence of this.

The industrial benchmarking uses a questionnaire to ask questions in various areas of your enterprise. The final indicators will be analyzed according to the standards of best practical performance of other companies that have completed the same benchmarking process. The industrial benchmarking should not be viewed as an audit, but rather as a process that helps the company to verify your position before world leaders in its industry, identifying their potential areas of opportunities for improvement and strengths. If the notes of each of the indicators actually reflect the current situation of the company, actual improvement opportunities will be found. A high score in a window that does not match the reality may prevent the company to work on the issue in order to resolve or improve it.

Step 2 - Meet with the benchmarking team in the middle of the day to get a consensus of indicators, where possible, which reflect the company's areas covered by the questionnaire.

The purpose of this meeting is to discuss team notes the questionnaire and define a single score for each indicator questionnaire according to the current reality of the company. The team leader must help the debate and, where possible, allow the consensus indicator of note. It is important to note, given the nature of a multilevel and multifunctional team, there must be a difference of opinion. This does not mean that someone is right or wrong. Any difference in indicators is a very important part of the process and can highlight individual areas for discussion.

It is noteworthy that the file answers of the industrial benchmarking questionnaire should be sent by the benchmarking facilitating leader via email, even before the visit of the company (before the 3rd step). In the case of a multi-application (various operating units of the same company), the questionnaire response should reflect the operation in which the manager and his team work, that is to think of each unit as a part of a bigger picture.

Step 3 - Visit of benchmarking facilitator; within a week or two, the benchmarking of the facilitator will spend two days in the company with the benchmarking team to bring an independent and external perspective. Going over the questionnaire on hand for discussion with the facilitator (consensus of notes). The company's data is compared to the global database and the facilitator prepare a report for the company, containing the results.

There must be a discussion during this meeting between the facilitator and the benchmarking team in which they will address each of the indicators in order to come to a consensus and adjust the scores defined in the questionnaire. The final marks should reflect the company's current situation in the different areas evaluated. It is recommended, where possible, that the visit to the company takes place within 1 or 2 weeks after the meeting of the benchmarking team.

On the first day of activities the industrial benchmarking facilitator must visit the production unit that will be evaluated. Later, there should be a meeting between the facilitator and the company's benchmarking team to discuss the questions in the questionnaire. It is necessary to carry out the work, with the assistance of a computer and a multimedia projector. When programming in the morning (Chart 2), i.e., during processing of the data, it is not necessary for the team or the leader follow the facilitator

Step 4 - Action Plan (optional) if the company is interested, the benchmarking facilitator can help develop an action plan for improvement opportunities raised by the results of industrial benchmarking.

It is recommended to prepare an Action Plan based upon the benchmarking results, in which the raised improvement opportunities can be analyzed in more depth. The third day of application of industrial benchmarking is a daily Action Plan, which consists of a workshop prioritization and preparation of the action plan for improvement opportunities verified by benchmarking. Responsibilities for actions and defined their terms are indicated.

Chart 2 Industrial Benchmarking Daily Schedule Outline

	1st day Programming
8h30 – 9h00	Meeting with the leader of the benchmarking team.
9h00 – 10h30	Visit to the factory of the company that is being reviewed by benchmarking and was scored in the questionnaire, together with the benchmarking team leader or manager of the company.
10h30 – 12h30	Meeting between the facilitator of industrial benchmarking and the company's benchmarking team to score consensus and setting defined in the questionnaire. The place of the meeting shall have a meetings table of compatible size with the number of people involved in the activity
12h30 – 13h30	Lunch
13h30 – 18h00	Continuation of the meeting between the facilitator of industrial benchmarking and the company's benchmarking team to score consensus and setting defined in the questionnaire.

To this day a flipchart, PC and projector, plus a large enough room for 10 people, with large table and chairs furniture it is necessary. The Action Plan should involve all the people who accompanied the industrial application of benchmarking in the first two days.

The three steps (after the visit to the plant) and four may be outside the company, in places such as hotels, convention centers etc., since there is evidence that indicates that other facilities can ensure smooth running of the work.

3.1.2 Benchmarking structure

This item shows the presentation structure of the benchmarking application model results, detailing the graphics and style tables and categories of comparative analysis. Categories are used for both the individual report of a company, for a sample, and is also used to identify opportunities for improvement. Therefore, it is important to expose the logic underlying the analysis.

The positioning of the company in relation to practical performance.

The practice and performance charts facilitate the study among international companies in its sector, according to the general practice and performance index obtained using the developed benchmarking model. In the following figure 19, Seibel (2001) describes the x-axis represents the overall rate of world class practices installed in the company and the vertical axis represents the performance index obtained.

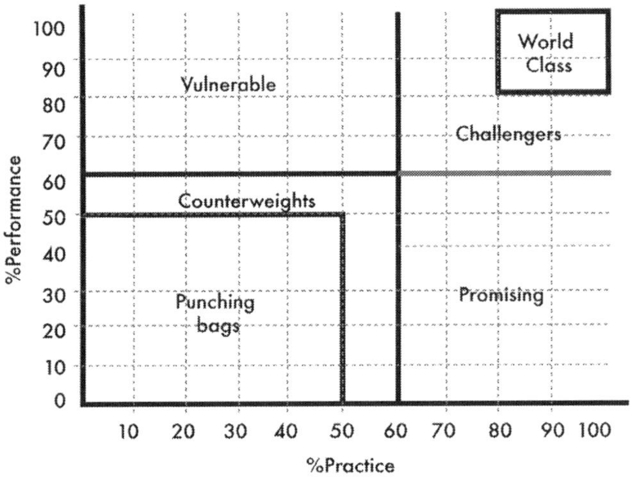

Figure 29 Boxing Analogy (Seibel, 2001)

The scale ranges from 0% to 100%. The position of a company is defined by the answers to the issues from questionnaire indicators, from which the general practice and performance indexes are calculated. Practice is a set of management and technological tools implemented in the company. E.g.: employee participation, automation and quality system. Performance are the measurable results of the enterprise. I.e.: inventory turnover, customer satisfaction and defect rate.

3.1.3 Company classification method: Boxing analogy

An analogy between the ability and performance used in boxing; a very popular sport in England; has been developed in order to analyze the company's positioning in the study according to the levels of practical performance and its corresponding ability to respond to competitive challenges. This analogy is based on benchmarking studies in the production world-class system developed in Europe by London Business School (LBS), in cooperation with the company's advisory group IBM England (SEIBEL, 2004). The company receives a name as its position in the practical performance diagram: World Class, Challengers, Promising, Vulnerable, counterweights and bag Showers (SEIBEL, 2001). The positions and their designations are shown in Figure 29.

World Class Company is defined in the model as one that has reached a level of practical performance and no less than 80% of standard as described world class. They are characterized by having implemented most of the best practices available in the industry and they are highly competitive in the international market.

Challengers are companies that obtained between 60% and 80% in the practical performance indexes, but have not reached the World Class level. Companies classified as Promising present higher educational practices at 60% and levels even lower performances to 60%. These are companies that have invested in the modernization and adoption of best practices, but have not yet obtained the proper return. The challenge of these companies is to improve their performance through the effective use of resources installed. Two situations are possible: the practices have recently been implemented and are currently on a learning curve; the practices were implemented however operational difficulties persist, caused by employee´s poor training or incorrect adaptation to the reality of specific process.

There are companies that although getting satisfactory results, have very low rate practices. These were classified as Vulnerable because the results are not consistent and its position is very unstable and difficult to sustain in the long run if the conditions of competition come to continue. Some companies that have very simple processes can achieve far superior results with the implementation of best practices. In all cases, it is necessary to analyze whether the high performance is being achieved by simplicity of the process or if the company is generating costs to offset operational inefficiencies, which reduces productivity.

A typical example of this I with companies that, dispatch requests by air, to compensate for delays in the lead time of the application in order to meet the deadline given to the customer; another example are companies operating with high levels of internal defects, but use rigorous inspection at the end of the process to prevent defective products from reaching the market. In both cases, the company pays for the costs of operational inefficiency.

In the Counterweights category we find companies with practical performance indexes between 50% and 60%. These companies are well below the industrial excellence. The group of lowest score was called "Bag Showers"; the companies in this group have scored less than 50% in practical performance. Their situation is serious, and their survival is threatened in the open market economy.

Processes Ratings are as follows: Performance Evaluation - PE; Product Development - PD; Plant and Equipment - PE; Investment - IV; Innovation - IN; Engineering Operations - EO; Environment, Health and Safety - EH; Quality - Q; Organization and Culture - OC; Production Cycle Time - CT (Figure 30 and Table 2)

In the application of industrial benchmarking in Brazilian companies, by facilitating the IEL / SC (2011) they were using, for example, 52 issues in their standard questionnaire based on the three pillars that are evaluated: Production, Stakeholders and products (Figure 31). For the IEL / SC, the company's strategy must be related to the speed of learning, seek the quality of market information and establish a performance indicator system to compare with other companies in benchmarking processes in a systematic way.

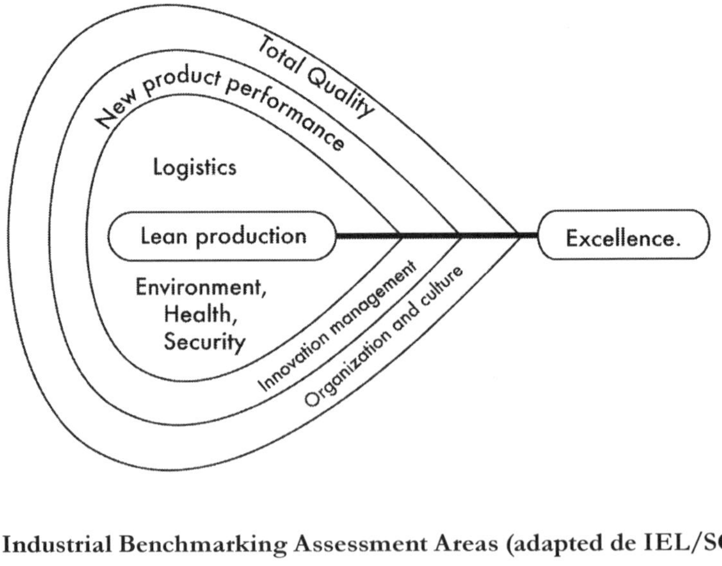

Figure 30 Industrial Benchmarking Assessment Areas (adapted de IEL/SC, 2006)

Table 2 Process Indicators (adapted from IEL/SC, 2006)

	INDICATORS	PROCESSES
AD 1	Customer satisfaction level	Total quality
AD 2	Market share	General Performance Indicators
AD 3	Employee satisfaction Level	Organization and Culture
AD 4	Inventory Turnover	Lean Production
AD 5	Cash flow	General Performance Indicators
AD 6	Productivity	Quality/production
AD 7	Product cost	Lean Production
AD 8	Performance measurements	Quality/production
DP 1	Field feedback	New product performance
DP 2	Production design, use and disposal	New product performance
DP 3	Concept time for production release	New product performance
DP 4	Market release production time	New product performance
DP 5	Functional Product Performance	New product performance
FE 1	Equipment layout	Lean Production
FE 2	Automation	Lean Production
FE 3	Pull Production	Production/Logistics
FE 4	Lot size	Lean Production
FE 5	Warehouse truck	Lean Production
FE 6	Issuance of product orders	Logistics
FE 7	Information Systems	Logistics
FE 8	Maintenance	Lean Production
FE 9	Production order priority change	Logistics
FE 10	Housekeeping/ 5 S	Lean Production
IV 1	Average age of plant and equipment	Lean Production
IV 2	Capacity investment	Lean Production
IV 3	Investment return time	Lean Production
IV 4	Technologic modernization investment	Lean Production
IN 1	Generation of innovative product concepts	New product performance

IN 2	Product life cycle planning	New product performance
IN 3	Product technology strategy	Innovation management
IN 4	Innovation capacity	Innovation management
IN 5	New product introduction (last two years)	New product performance / Innovation management
IN 6	University and research center relationships	Innovation management
IN 7	P&D infrastructure	Innovation management
IN 8	Strategy to obtain new technologies	Innovation management
OE 1	Concurrency in engineering	New product performance
OE 2	Multifunctional involvement (internal)	New product performance
OE 3	External involvement	New product performance
OE 4	Change Control	New product performance
OE 5	Tools support introducing new products	New product performance
OE 6	Automated project development	New product performance
MS 1	EHS Policy	Environment, Health and Safety
MS 2	Management systems	Environment, Health and Safety
MS 3	Risk assessment	Environment, Health and Safety
MS 4	External communication	Environment, Health and Safety
MS 5	Pollution control	Environment, Health and Safety
MS 6	Frequency of reported environmental incidents	Environment, Health and Safety
MS 7	Waste	Environment, Health and Safety
MS 8	Total reduction of waste	Environment, Health and Safety
MS 9	Accident investigation	Environment, Health and Safety
MS 10	Damage sustained by employees	Environment, Health and Safety
MS 11	Occupational health problems	Environment, Health and Safety
MS 12	Sick leave (days/employees in the year)	Environment, Health and Safety
Q 1	Quality of vision	Total quality
Q 2	Quality models and procedures	Total quality
Q 3	Process capability	Total quality
Q 4	Product reliability in service	Total quality

Q 5	Warranty costs	Total quality
Q 6	Defects (internal)	Total quality
Q 7	Deliveries percentage within term	Logistics
Q 8	Relationship with Suppliers	Total quality/Lean Production /Logistics
Q 9	Initial production quality of a new product	Total quality / New product performance
Q 10	Scrap costs, rework, recycling	Total quality
OC 1	Vision of the future	Organization and Culture
OC 2	Sharing vision, mission goals	Organization and Culture
OC 3	Production strategy	Organization and Culture
OC 4	Leadership style	Organization and Culture
OC 5	Involvement of employees	Organization and Culture / Total quality
OC 6	Flexibility at work	Lean Production/Organization & Culture
OC 7	Systematic use of diagnosis	Organization and Culture / Total quality
OC 8	Staff development	Organization and Culture/Innovation management
OC 9	Customer orientation	Total quality / Organization and Culture
OC 10	Tools for troubleshooting	Total quality
OC 11	Innovative environment	Organization and Culture / Innovation management
TC 1	Company's total cycle time	Logistics / Lean Production
TC 2	Material processing cycle time	Lean Production
TC 3	Customer Order Processing Time	Logistics
TC 4	Delivery performance of suppliers	Logistics
TC 5	New product/process introduction time	New product performance / Logistics
TC 6	Equipment preparation time	Lean Production

According to Schuch and Castro (2011), In a study conducted with the database of international industrial benchmarking, with 905 industries evaluated in 35 countries (such as USA, India, Australia and European Union countries), the adoption of practices related to Lean production, the structuring of a product development process and promoting organizational learning are cited as the main strategies to gain competitiveness of the best companies in the world, as an example, there is the Volvo do Brazil, winner of the National Award for quality in 2009 and considered one of the best companies to work for, the company implemented a number of improvements that have gone through the adoption of a model of management, benchmarking of use for the external reference of the main indicators, greater alignment of indicators with the company's strategy and adoption of a model of learning and intellectual capital management.

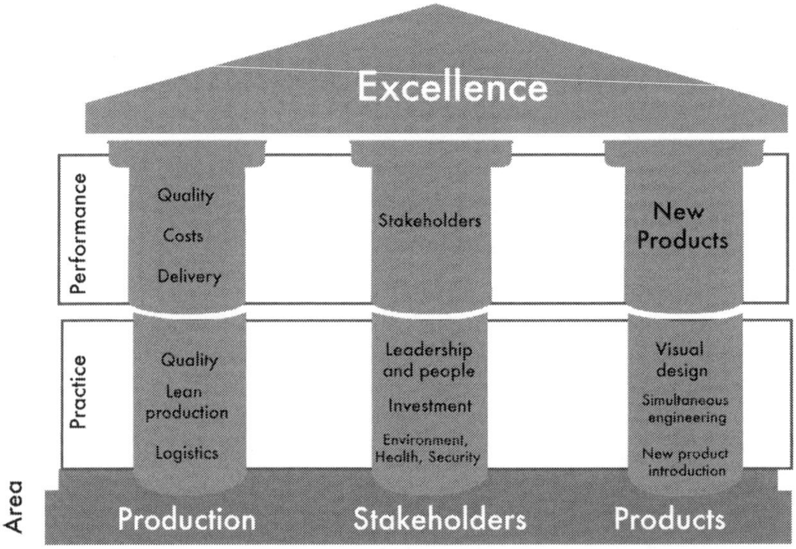

Figure 31 Evaluation of areas of industrial benchmarking (IEL/SC, 2011)

3.2 Grey system theory: selection and evaluation method

When searching for new ways to select and evaluate the indicators of the Enterprise Diagnostic Method (EDM) resulted in the knowledge of Grey theory. The definition of the Grey System is consistent with the concept of the black box, knowledge contained in the system, which are called unknown and a Grey system. A Grey matrix with some of their properties (for example, a number of rows and columns) known but some unknown factors are called a grey system.

Professor Deng Ju-Long (1982) describes the Grey System term which is also called: Grey Relational Analysis - GRA and Gray Correlation Analysis. There are many articles which employ Gray name and not the initial description, Professor Deng Grey gave to the Grey System. In the research and application work, we adopt the term Grey system that can also be described and translated into a correlated analysis method of Grey and Grey system.

Authors have adopted a hybrid method of two attractive techniques, i.e., the effective fusion of hierarchical-analytical process (AHP - Analytical Hierarchy Process) and the gray system to better the decision-making processes.

AHP selects the criteria's importance and the grey system selects the scores. The AHP developed by Saaty (1980) is a useful approach to solve complex problems, while the grey system is used to explore the qualitative and quantitative relationships between abstract and complex succession and capture the dynamic characteristics during the development process.

Research carried out in March 20, 2011, in Compendex database, in which about 290 articles were surveyed, from 2007 to 2011 (last five years), related to the application of the term Grey, shows that after applying the Pareto principle 80/20 (Graph 3) to such methods the main result, according to the keywords is knowledge applied mostly in: systems theory; Intelligent systems; forecast

Demand forecasting; Mathematical models; Neural network; Decision Making; optimization; Quality control; competitiveness; Hierarchical systems; cybernetic; Supply Chain; Decision theory; Product development; Project Management; Production Engineering; industry; Product design; Customer Satisfaction; Computer simulation; Industrial engineering; economy; E-commerce; Management Supply Chain; Problem Solving; technology; Fuzzy systems; Quality assurance; project; sales; Agricultural products; Cluster analysis; Analytical Hierarchy Process; Granular materials; algorithms; Imaging Systems, among many other applications.

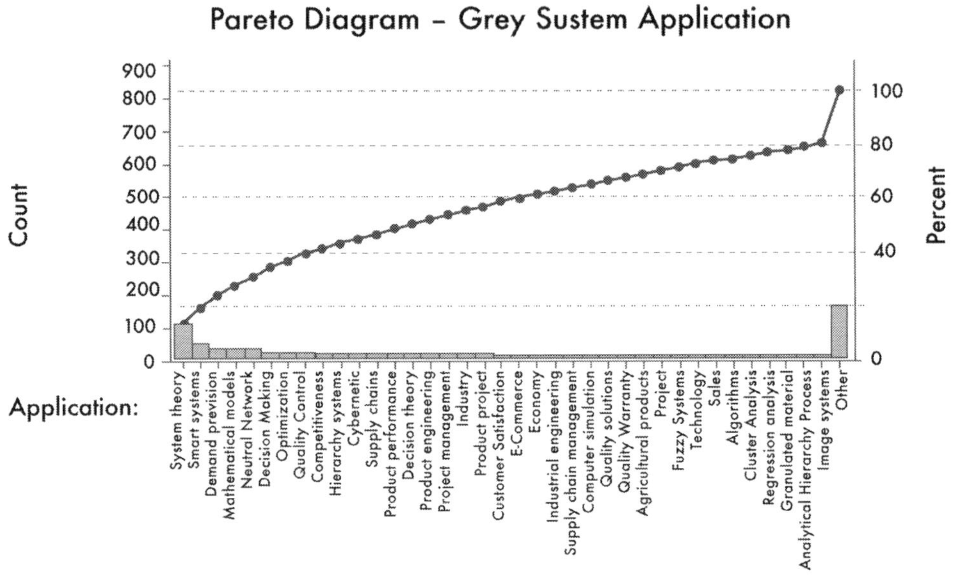

Graph 3 Applying the Grey mapping system (Compendex base, 2011)

According to the same cutoff criteria, according to the Pareto principle 80/20, there were classified areas that are applied to the knowledge of these two methods most often: Systems Science, Mathematics, Control System, Statistics Mathematics, Business, Industrial engineering and management, technology information, probability theory, industrial economics, biomedical engineering, artificial intelligence and other various applications.

3.2.1 Grey system application

The Second publishing from Tie-Jun Sha (2008), states that Gray correlation analysis is a statistical analysis of multi-factors. According to the sample data of all the factors, the degree of Grey correlation is used to describe the strength and weakness, size and shape of the relationship between factors. If the sample data reflects the same mutative situation of two factors, the degree of correlation is higher, the opposite results in a lower correlation. As an example the combination of these three designs (A1, A2, A3) together with the cutting method provides unique design and position (A4), a total of four designs are examined in Table 5. In accordance with Tie-Jun and Sha (2008) this is data comprising decades of experience in implementing several projects that generated the technical indexes of economy of various designs and these results are shown in table 3.

Table 3 Technical Index economy to evaluate design improvements

(TIE-JUN; SHA, 2009)

Index / **Project**	**A1**	**A2**	**A3**	**A4**
X1: Accuracy (%)	90	95	99	99
X2: Structure Investment (x $ 10.000)	1	5	100	0.1
X3: Labor Cost (x $ 10,000)	30	9	9	100
X4: Count rate (number per hour)	2000	1200	60000	500
X5: Area size to cover	major	large	major	largest
X6: Reconstruction difficulty degree and ease	Common	Difficult	More difficult	Easier

The indices in Table 3 show a semi-quantitative result. Then they take up some form of estimated numbers to quantify X5 and X6 which is expressed in table 6 and reorganizes all indexes quantitatively.

Table 4 Quantity Evaluation Index

Project \ Index	X1	X2	X3	X4	X5	X6
A1	90	1	30	2000	3.75	5
A2	95	5	9	1200	2.5	1.25
A3	99	100	9	60000	6.25	3.75
A4	99	0.1	100	500	1.25	8.75

Table 4 is arranged as a non-dimensional form by linear standardized method, while all indexes are unified and positive:

$$\mathbf{X} = \begin{bmatrix} 0.909 & 0.100 & 0.300 & 0.033 & 0.600 & 0.571 \\ 0.960 & 0.020 & 1.000 & 0.020 & 0.400 & 0.143 \\ 1.000 & 0.001 & 1.000 & 1.000 & 1.000 & 0.429 \\ 1.000 & 1.000 & 0.090 & 0.008 & 0.200 & 1{,}000 \end{bmatrix}$$

At this time, the largest optimal proportion of samples X0 = (1, 1, 1, 1, 1, 1), due to the formula (1) calculates the absolute difference of the sample X0 and Xi is the absolute difference Δ matrix.

$$\Delta_{ij} = |X_{0j} - X_{ij}| \quad (i = 1,2,3,4; \quad j = 1,2,3,4,5,6) \qquad (1)$$

$$\mathbf{\Delta ij} = \begin{bmatrix} 0.091 & 0.900 & 0.700 & 0.967 & 0.400 & 0.429 \\ 0.040 & 0.980 & 0.000 & 0.980 & 0.600 & 0.857 \\ 0.000 & 0.999 & 0.000 & 0.000 & 0.000 & 0.571 \\ 0.000 & 0.000 & 0.910 & 0.992 & 0.800 & 0.000 \end{bmatrix}$$

The definition of the weight is required once the Δ (max.) And Δ (min.) are calculated. In this case Tie-jun and Sha (2009) describes the different weights to the indices (see table 8 indicators x1, x2, x3, x4, x5 and x6). These are given different weights due to the judgment of importance. The weight represents the degree of importance of information and is defined between zero and one, as a variable which belongs to the real numbers in the interval (0, 1) so that the sum of the weights equals 1 (100%).

Here Δ (max.) = 0999Δ (min.) = 0, referred to the weight of the various indexes (Wj): 0.2, 0.2, 0.2, 0.1, 0.15, 0.15. If the calculated correlation coefficient; do the following processing of data in the absolute difference matrix.

$$\varepsilon_{ij} = \frac{\Delta(\min) + \rho\Delta(\max)}{\Delta_{ij} + \rho\Delta(\max)} \qquad (2)$$

The coefficient distinguished and takes the value between 0.1 and 0.5, ρ assigns to the value equal to 0.3, to calculate the correlation coefficient matrix:

$$\varepsilon_{ij} = \begin{bmatrix} 0.767 & 0.250 & 0.300 & 0.237 & 0.428 & 0.412 \\ 0.881 & 0.234 & 1.000 & 0.234 & 0.333 & 0.259 \\ 1.000 & 0.231 & 1.000 & 1.000 & 1.000 & 0.344 \\ 1.000 & 1.000 & 0.248 & 0.232 & 0.273 & 1.000 \end{bmatrix}$$

Other authors attribute to the ρ value of 0.1 (Chan and Tong, 2007) and Chan (quoted Deng, 1989) states that the value of 0.5 is usually applied.

To calculate the correlation applies to equation (3).

$$r_i = \sum_{j=1}^{6} w_j \varepsilon_{ij} \qquad (3)$$

$$r_i = \begin{bmatrix} 0.153 & 0.050 & 0.060 & 0.024 & 0.064 & 0.062 \\ 0.176 & 0.047 & 0.200 & 0.023 & 0.050 & 0.039 \\ 0.200 & 0.046 & 0.200 & 0.100 & 0.150 & 0.052 \\ 0.200 & 0.200 & 0.050 & 0.023 & 0.041 & 0.150 \end{bmatrix}$$

Calculated r1 = 0.4048, r2 = 0.5315, R3 = 0.6941, R 4 = 0.6624, the project evaluation assembly sequence is: A3> A4> A2> A1. This shows that the project A3 is the best plan. According to the Grey association, in the evaluation of the six indicators, it can be known that "the use of unification and of the same capacity material box" is the best option.

3.3 Introduction to Hoshin planning

The term Hoshin Kanri (Japanese), according Akao (1997), has many meanings and the most common terms are: Management by Guidelines, Hoshin Planning and Deployment Guidelines. King (1989) uses the term Hoshin planning, Akao (1997) uses the term Deployment Guidelines, Campos (2004c) uses the term Management by Guidelines, James Womack in the preface to the English edition of "getting the right things done" of Dennis (2006), translates the term Policy Deployment (hoshin kanri) as Deployment Policy. In publications in English, Akao (1991), Cudney (2009) and Jackson (2006) use the term Hoshin kanri, although describe the translation into English as policy deployment. This book adopts the term Hoshin planning or Hoshin kanri based on King (1989).

The word policy is used as a translation for the Hoshin means and means pointing the direction (ho - direction, shin - needle), deployment is used as a translation for Kanri and is the means of administration or control (kan - control, ri - logic / reason), that is, one can understand the kanri hoshin expression as true north, established within the organization and serves as a compass or guide for the strategic actions of the company.

Hoshin Kanri is a strategic decision-making tool that is geared to meet the organizational objectives so aligned with specific plans at all levels of the organizational structure. Jackson (2006) suggests not using the Hoshin simply to deploy the strategy, but use the Hoshin to manage the business and build a Value Stream according to the Lean approach.

Cudney (2009) believes in the Lean approach as a solution for businesses. For him the goals should be integrated strategically therefore presents the Hoshin kanri as a method for implementing the breakdown of targets. Lean principles are important not only for continuous improvements, but also for the development and maintenance of the company's competitive strategy.

The strategic vision of the company considers the Lean initiatives as a means to achieve the strategic objectives of all areas of the company and therefore to obtain good results it is essential to apply the methods and tools integrated with the Hoshin Kanri. Cudney adds to its implementation model of Hoshin planning the integrated application of mapping the value stream (Figure 32).

The Lean Enterprise integrates all levels of the organization, as it focuses on process improvement and overall perspective (Murman et al., 2010). The company needs to implement continuous improvements with holistic and integrated to achieve your goals.

Hoshin Kanri began in Japan in the 1960s, as well as the statistical process control (SPC) became known and applicable by the Total quality control (TQC). TQC employed by Dr. Armand Feigenbaum is an administrative model fitted by the quality control research group JUSE (Union of Japanese Scientists and Engineers). In Brazil TQC has been successfully implemented and contributed in changing culture and success of numerous companies such as Petrobras, Multibrás, Embraco and Usiminas (CAMPOS, 2004b).

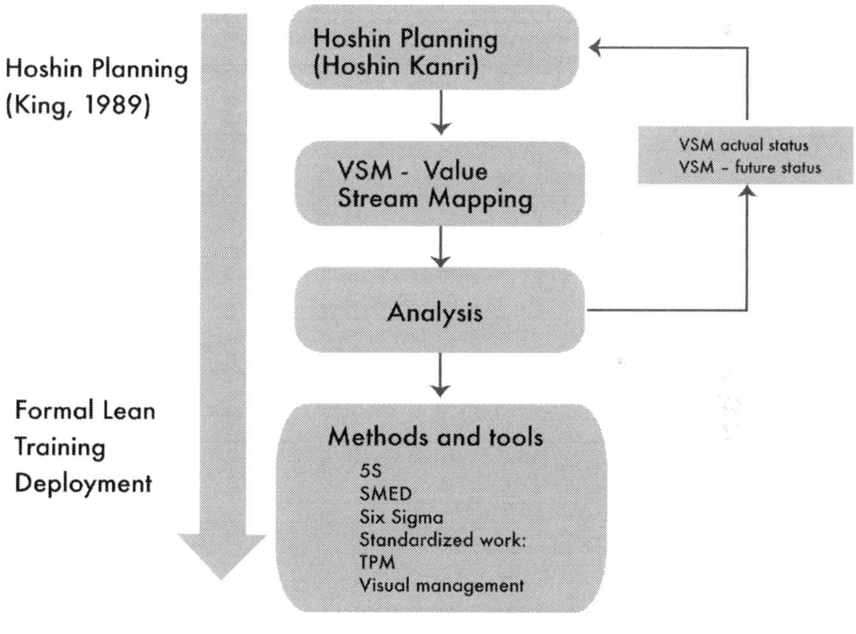

Figure 32 Improvement strategy (Adapted from CUDNEY, 2009)

The business strategy system

Thus shown in Figure 33 is a business strategy system, taking into account the four basic dimensions of Total quality (safety, quality, delivery and cost). To meet customer needs, listen to voice them and to ensure greater and better satisfaction to implement the methods and tools, but always aligned to the Hoshin planning. The application of Lean tools and methods becomes the mean to achieve the objectives, focused on continuous improvement, waste elimination and customer appreciation.

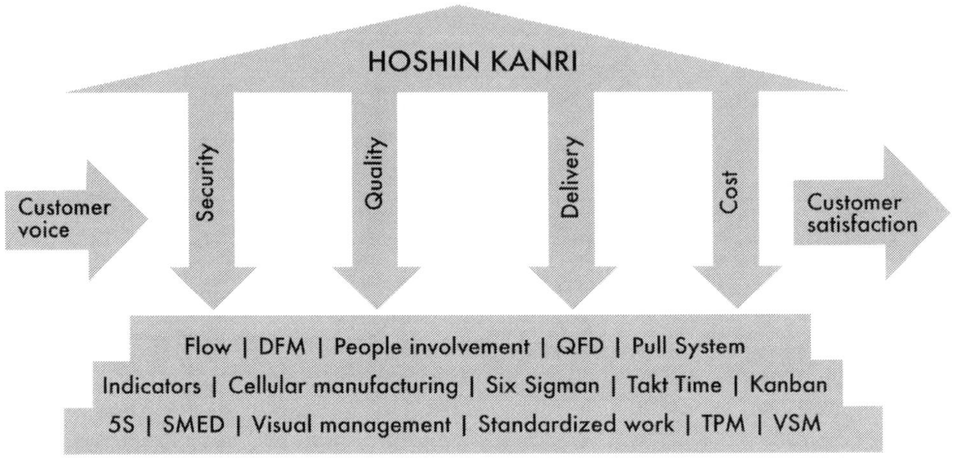

Figure 33 Business Strategy System (CUDNEY, 2009)

Safety concerns the welfare of employees and users of security products and / or services. The quality is directly linked to the satisfaction of internal customers, external and depends on the management of routine day-to-day processes. Delivery cares to serve customers at the right time, right place at the right quantity and in this dimension of Total quality, a measure of the product is delivered or final and intermediate services of the company and no less important than the cost is concerned with the care of a final and intermediate value compatible with the business expectations (CAMPOS, 2004b). The business strategy system is related to the need for synergy of people and processes to achieve a more sustainable business model, global and competitive.

The Hoshin Kanri is a change management system for critical business processes. This is a method to improve process performance and achieve strategic objectives. Kanri Hoshin can integrate with the focus, responsibility, involvement and participation of all levels of the organization. This synergy fosters innovation, internal organization and the approach becomes the improvement projects that add value and not just an execution of tasks. The voice of the customer must be heard and strive to improve the quality dimensions, delivery, safety and cost continuously.

Catchball and the Hoshin kanri

The catchball refers to the give and take needed between the various levels of management during the planning process. Through catchball, strategies (long-term plan) static (short-term plan) proliferate in the organization (KONDO, 1998).

The catchball seeks to link the vision of the direction with the daily activities of the shop floor team members; Dennis (2008) describes the following steps:

- Management will develop a vision of what the organization needs to do and skills that need to be developed. "You play" this vision to senior managers.

- Senior managers "take" the vision of the direction and turn it into hoshins. Then "play-in" back to the direction and ask, in practice, "That's what you talk about? These activities make our vision a reality?"

- The direction provides feedback and guidance to senior managers. The hoshins can go back and forth several times.

Over time, you arrive at a consensus. The direction and senior managers agree that "these are the hoshins that our company will use to reach our vision."

- Senior managers play their hoshins for middle managers to catch and turn them into activities. These, in turn, are thrown back to senior managers who provide feedback and guidance. Over time, you arrive at a consensus. The senior and mid-level managers agree that "these are the activities that is used to meet the hoshins senior management and, in turn, will reach our vision."

- The middle managers, in turn, will play their hoshins his subordinates. The process culminates with the objectives performed by members of individual teams.

The catchball requires a different kind of leadership and mutual respect basis. The leader relies on knowledge of the team members and their ability to achieve the desired result. The team members trust the leader's judgment to select the focus area, but is encouraged to respond with facts. The catchball is a refining process that helps management and staff understand what is real. The catchball is a political process in the best sense of the term, but the policy is anchored in data.

The catchball system implies frank discussions, based on reality, between the various levels of management and also within the same. Dennis (2006) describes the usage the catchball for better strategic deployment in the Atlas company to achieve their goals in manufacturing and product development.

The result of Hoshin planning is shown in a tree of activities aligned with strategic and philosophical goals of the business (Figure 34). From top to bottom, in the figure, the more concrete business goals, usually involved with financial measures and expressed more widely in short phrase that expresses the vision, direction and values as an example: 252 million revenue this year and operating profit of 5% compared to last year.

In the second level of the tree performance objectives can be assigned, such as the delivery and return, and it is recommended according to the philosophy of Lean Production which adopts the improved use of tools and management reporting A3.

In the third level of the tree are generated action plans at the departmental level where also recommends using the A3 tool. These plans may need to kaizen workshop or even specific projects with use of specific tools of Six Sigma, Lean Production, project management, quality tools etc.

The catchball should be aligned to Nemawashi method, which means "prepare the tree to be transplanted" and connotes the consensus-building process that creates an alignment. Nemawashi involves the review of a Hoshin with all affected customers before implementation. Therefore, planning usually takes longer, but its implementation is more effective. Hoshin's clients are those who will use the plan, who will be affected by the plan, those who approve the plan and to be able to improve the plan. The more reviews based on customer feedback the fewer surprises will arise in the senior management plan, as Witcher (2002) and Dennis (2008).

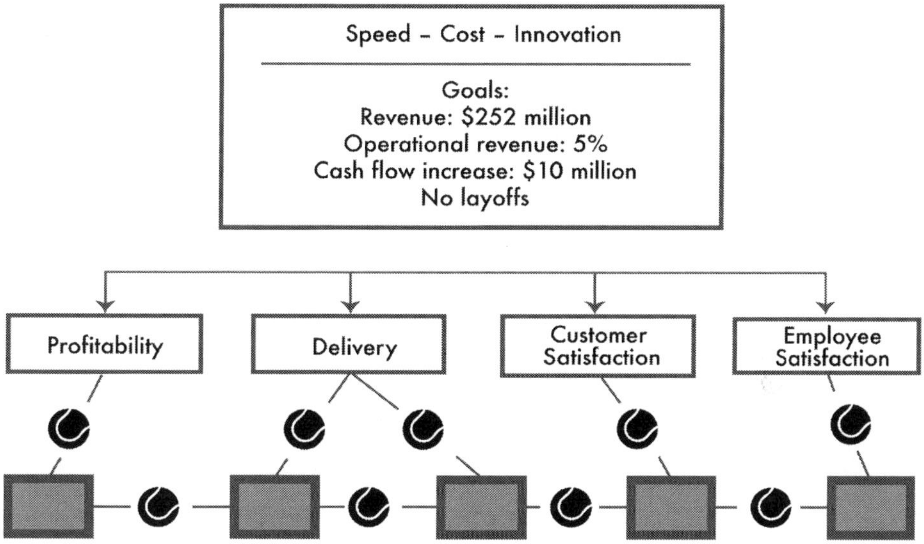

Figure 34 Illustration of catchball (DENNIS, 2006)

The performance objectives of companies such as productivity, quality, cost and safety in general require multifunctional goals, group work where there are representatives from departments such as engineering, production, maintenance, quality control of production, therefore, coordination may be performed by a control department to lead the hoshin planning process for quality, lead the establishment of goals and means by nemawashi and catchball apply the PDCA in micro and annual level to confirm success. This coordination makes them visible problems and supports the countermeasure activities.

3.3.1 Organizational need levels

The concept of man's basic needs is already known thus organizations also have hierarchies and needs, such Cudney (2009) describes five levels of hierarchical need for an organization that is directly related to the six steps for implementation the hoshin planning are: the company's vision, alignment, self-diagnosis, process management and the focus on targets (Figure 35).

Figure 35 Organizational need levels (CUDNEY, 2009)

A business plan Hoshin kanri to transform a company into world class enterprise begins with the organization's vision and second Womack (1998) it takes five years, Cudney (2009) states that development happens in 3-5 years. Table 4 describes the five levels of the hierarchy and their relationship with the organization's needs, as well as the Hoshin methods. To take as an example the forms for each of the methods listed in the table are exposed by Cudney (2009).

Table 5 of the organization need ratio, planning and Hoshin method (CUDNEY, 2009)

Organizational Needs	Steps to Hoshin planning	Hoshin method
Company vision	5 year vision 1 year plan	*Hoshin* strategic plan table
Alignment	Development	*Hoshin* plan table
Self-diagnose	Implementation	*Hoshin* action plan
Process management	Monthly review	*Hoshin* Implementation plan
Focus on target	Yearly review	*Hoshin* Implementation review

3.3.2 Hoshin process planning

Cudney (2009) proposes six steps to implement Hoshin planning. The first step in the implementation of Hoshin planning is the first step when the company develops the vision for the next five years, defining strategic goals, based on internal, external environment and the desired changes. In this context, goals, capabilities and culture for the future of the business are defined. This view is a communication tool, like a compass, guides decisions into the future, decisions that are reviewed every five years. Details can be clarified to see the form, frame of Hoshin strategic plan.

The second step; after the 5 year vision is already defined; is to develop the annual plan for the company with activities and continued improvement projects. The strategy is long term; therefore the goals are defined for the next 5 years. We evaluate the external factors: competitors, economic, political and social; in the search for competitiveness and sustainability of the business then the organization´s problems are analyzed. Management prioritizes objectives based on dimensions of total quality (safety, quality, delivery and cost) goals and measurements for Hoshin planning are defined as they unfold.

The third step of the Hoshin planning unfolds the annual plan of 1 year for all company departments with the activities and continuous improvement projects. The feasible goals are established for each department, according to the opportunities and needs for improvement for the company. At this point the departmental strategies are defined with managerial and tactical level.. Details can be clarified to see the form, frame Hoshin plan.

In the next step, each department must implement the activities and continuous improvement projects, since it is aligned to the annual plan and the strategic guidelines five years. At this stage, therefore, pulling the implementation of actions with the aid of methods and Lean tools also involves alignment with master plan and set goals and measures for management's strategy to tactical level. Auto diagnostic details can be clarified to see the form - Hoshin Action Plan.

For the fifth step and the management of activities, monthly, one must have a way to communicate the quantitative results to the company's management; so you can keep up with the continuous improvement, activities planned versus completed and its effects. When a goal is not achieved it is necessary to create new shares of corrections to recover the quantitative difference and achieve the expected results. It is the responsibility of each department manager to achieve the targets, as well as apply the problem-solving methods to analyze and solve existing ones. The management of monthly processes can be exemplified in the form Hoshin Plan Implementation.

The last of the six steps to the Hoshin planning is conducting an annual review of the strategies, monitoring the annual evolution of the company's results. The annual review creates new opportunities for business and ensures alignment with the five-year plan as well as the future vision for the business. Review of Hoshin implementation. The focus on goals and guidelines, through the windows can be clarified to see the form - Hoshin Implementation Review.

The strategic planning processes as well as the strategic management are the key to business success. The six steps to the Hoshin planning Cudney (2009) are just a simple plan, which deserves attention within the implementation of planning, the use of value stream mapping, which guides the application of methods and Lean tools as a business strategy to achieve sustainable results according to predefined view.

The PDCA cycle and the Hoshin planning process

Another case of application of Hoshin Kanri known and described by Jackson (2006), in which the PDCA (Plan, Do, Check, Act), along with the progression step-by-step in the planning process hoshin view Jackson are shown in figure 36 and 37. This planning includes the design of a business strategy or experience, and assigns responsibilities for planners of the four implementation teams. Planning script Hoshin kanri describes the strategy in seven steps.

The first team is usually a management team with strategic positions in business units and worries about the entire enterprise, growth, brand, product line, service, physical arrangement and the value stream. A Hoshin management team seeks aggregation flow value in the supply chain that increases the value of the company. Company staff can participate in the Hoshin teams, though; a cross-functional team usually produces better results and visions. After making the diagnosis, Hoshin team sets out a strategy for the company to turn. It is suggested as a business strategy and as an experimental project, consisting of the seven steps to solve the problem or challenge. In designing a strategy, the staff will utilize Hoshin as intended; to define the strategic elements, including the identification of the first seven steps as follows:

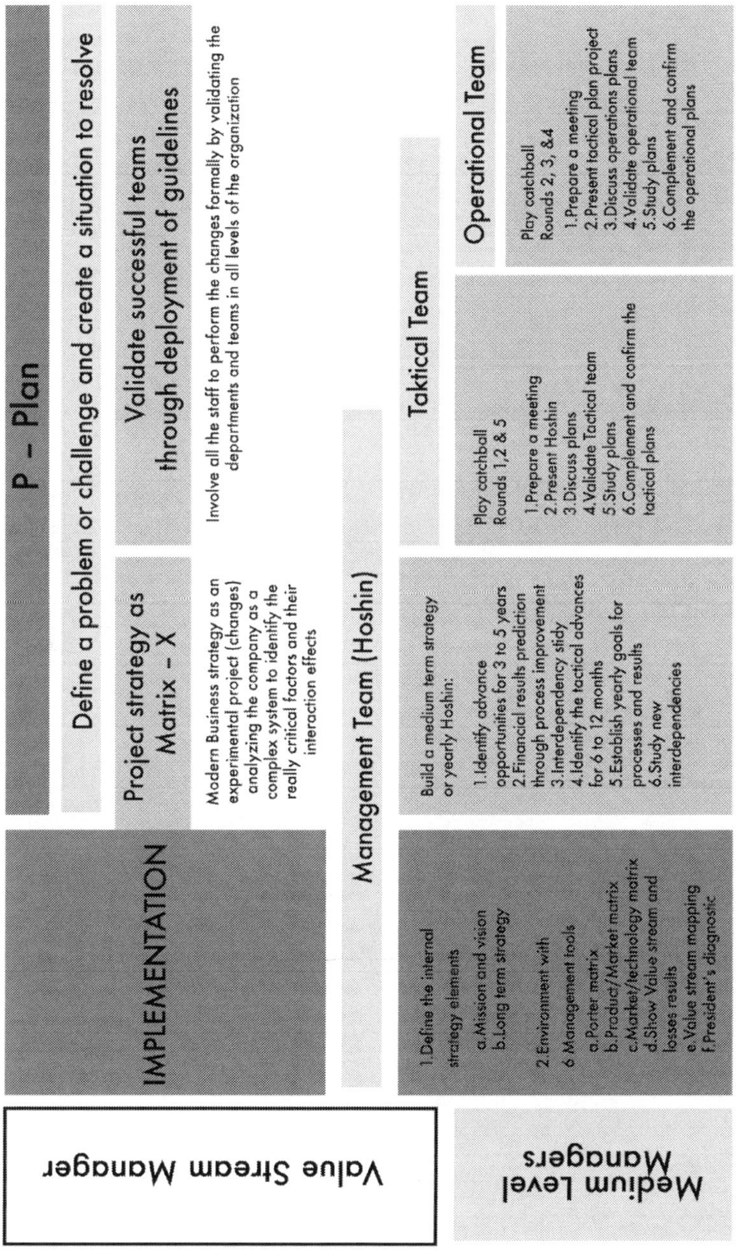

Figure 36 Hoshin Planning – PDCA (JACKSON, 2006)

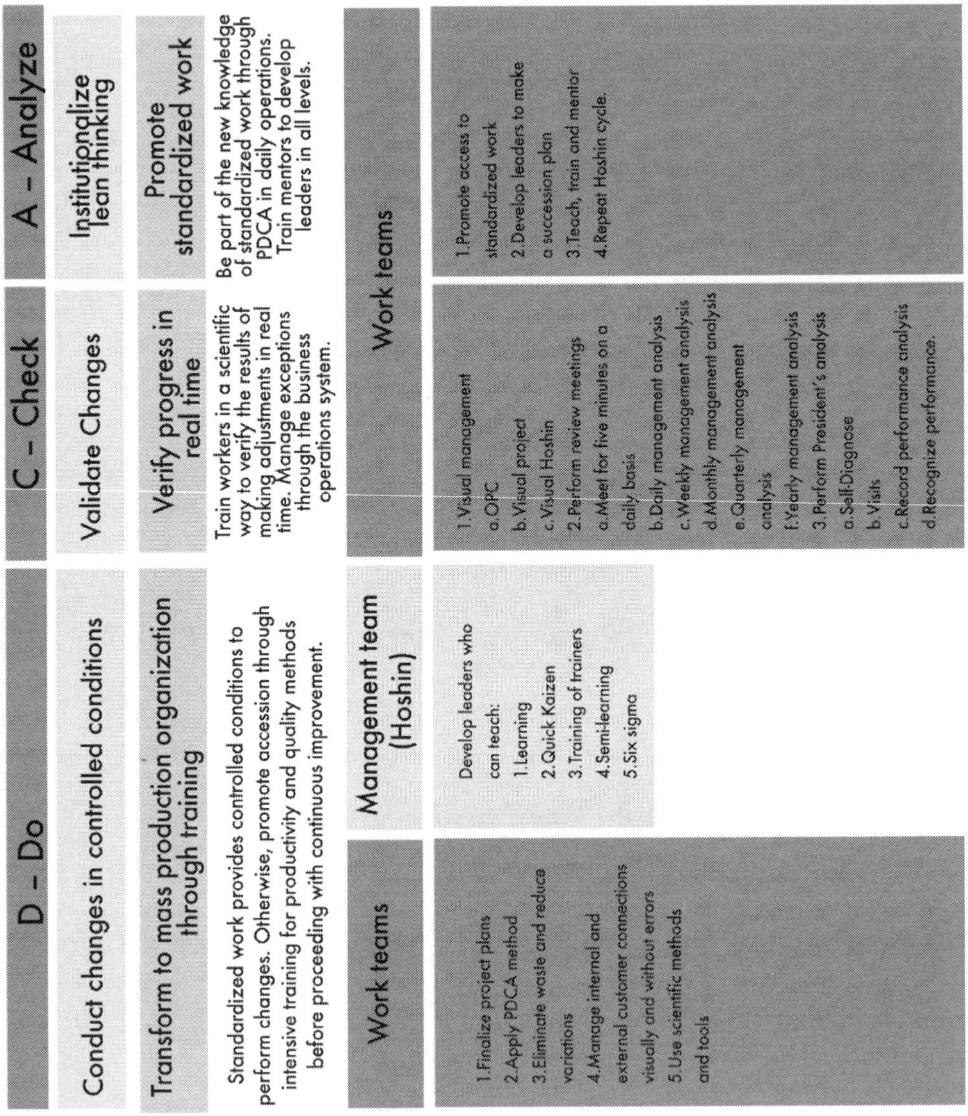

Figure 37 Do, check and standardize with Hoshin – PDCA (JACKSON, 2006)

3.3.3 Jackson's seven Hoshin planning steps

According to Jackson (2006), strategy is as a scientific hypothesis because no one can know the result of a previous strategy, especially a dynamic strategy that involves improving the way business is done and it is up to managers to struggle and find out what will happen. Thus, setting up seven steps, where, under controlled conditions of standardized work, the Hoshin process involves all managers and employs a hypothesis that through data and facts the company can establish its business strategy, operations, marketing, products and other areas.

The Hoshin planning steps are performed by teams (working groups) that eventually include the top-tier management, and other hierarchical levels to participate in the Hoshin process stage. Every step of the PDCA in Hoshin system has a different purpose, depending on its duration and its relation to the overall objectives of the organization. In general we can state that the longer the cycle the greater the level of responsibility for the management hierarchy. In addition, the process never ends Hoshin kanri. The strategic improvement cycle is repeated once a year. The companies that are starting their Lean or Six Sigma transformations can take up to 18 months to complete the first cycle, but you can execute the cycle twice a year to accelerate the learning organization and the Hoshin planning.

The seven steps of Hoshin planning represent actual work that teams need to do to implement the business strategy of experimental design to address the emerging challenges or problems (see Figure 38). Table 6 defines the four types of Hoshin team (teams) and seven kinds of PDCA cycles or steps that are synergistic and all aligned. The four teams have the following basic responsibilities:

Hoshin team has overall responsibility for the strategic planning and implementation process and denominations and guides the first three steps: (1) long-term strategy, (2) medium-term strategy, and (3) Annual Hoshin. These three steps typically have focused on improving business processes, in which all the demanding business needs a cross-functional coordination or coordination and integration between the company and its suppliers and customers.

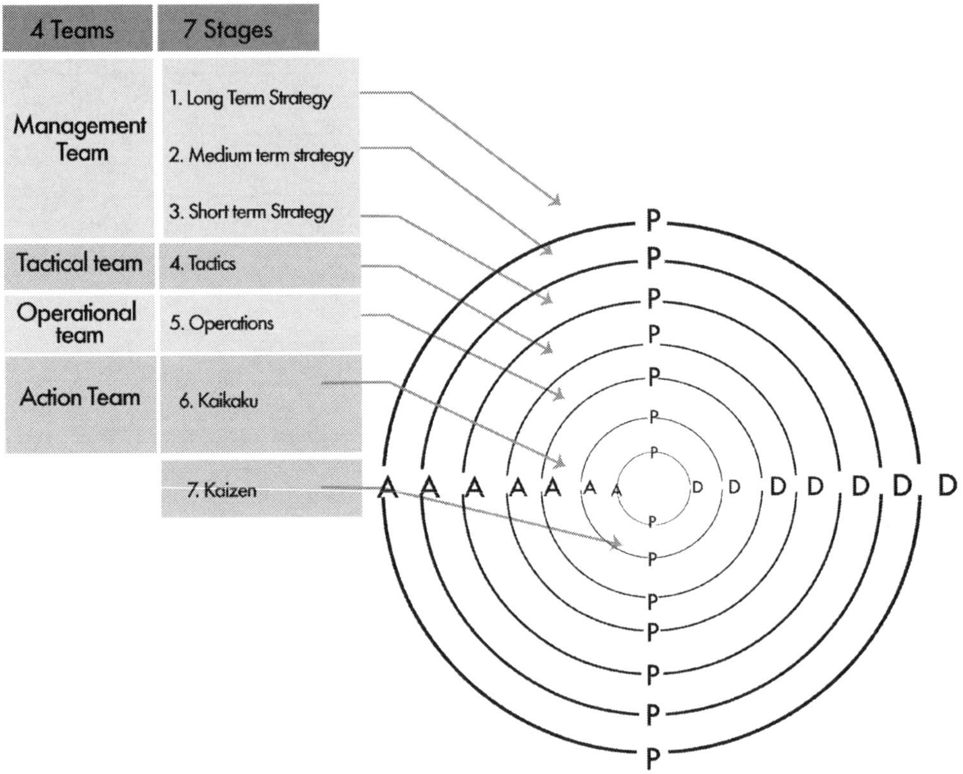

Figure 38 Times and stages of change (Jackson, 2006)

Tactical teams design and guide the four steps of Hoshin: tactical initiatives to develop specific competitive capabilities. Tactical initiatives usually have focused on functional improvement of business processes, i.e., the process of marketing, engineering, manufacturing, quality etc., but they also address all the significant elements of cross-functional coordination necessary for successful implementation and culture change.

Operational teams design and guide the 5th step of Hoshin planning, identifying operational projects to improvements in products and processes. Operational projects also focus on the improvement of business processes and are directed to a coordinating role. Operational teams are cross-functional, and are necessary for the successful implementation of Hoshin planning.

Table 6 The table teams and the seven Hoshin steps (JACKSON, 2006)

Four Times		Seven Hoshin Steps		
1	Time hoshin	1	Long-term strategy	A general plan of action aimed at the long-term period, planning 5-10 years to make major changes or adjustments in the mission and / or vision of the business.
		2	Medium-term strategy	A partially completed action plan includes financial targets and process improvement measures aimed at exceeding three to five years to develop the capabilities and align the trajectory of operations with long-term business strategy.
		3	*Hoshin annual*	A very concrete action plan aimed at a long-term, next 6-18 months to develop competitive skills and align the trajectory of trade, with the medium-term business strategy.
2	Time Tactic	4	Tactic	Concrete initiatives 6-18 months set by the annual Hoshin, undertakes to develop new specific skills, applying new technologies and methodologies in general business processes.
3	Operating Time	5	Operation	The concrete projects take three to six months, defined by annual Hoshin, undertakes to apply new technologies and Lean methodologies standardized processes in specific business functions.
4	Work Time	6	Kaikaku	The concrete projects take 1 week to 3 months, usually defined after the implementation of the annual Hoshin, pledging to apply the methods and Lean tools.
		7	Kaizen	Tools of quality and short-term problem solving to eliminate defects, errors and anomalies that may arise during the standard daily work and resulting improvements arising from employee suggestions.

The work teams proceed with the sixth and seventh steps in Hoshin. The sixth step is to implement periodic improvements of great magnitude called kaikaku and the seventh step is to implement incremental continuous improvement, called kaizen.

3.4 Hoshin Management Model Report

The A3 report is an effective communication tool. The use of A3 is recommended for Hoshin planning to integrate hoshins sectors and company, the A3 is also used for troubleshooting to summarize problems and countermeasures. The A3 is often used also to describe and show the current situation, where later one can offer new ideas and or improvements through the proposed A3. There is no single form for the A3, but can have a logical structure, containing basically: the Hoshin generation, provision Hoshin, implementation of Hoshin and final evaluation (JACKSON, 2006).

Sobek II and Smalley (2010) describe an example of proposed A3 report to implement a corporate credit card. In the case of A3 report (see Figure 39) the logical structure follows a stream that can identify and meet the historical fields, current condition, proposal analysis times and costs, implementation details and timeline (implementation schedule). This structure joins the intellectual and professional development to meet the company management needs. Thus, the A3 report system aims to be a simple problem-solving and visual management and is based on seven elements: logical reasoning process; objectivity; results and processes; synthesis, distillation and visualization; alignment; external coherence; and systemic point of view.

Not only is the A3 highly visual, it is also highly portable, so it facilitates the management walking around, a practice that encourages managers to discuss the progress in achieving the company's goals often face to face with their direct reports, and, with their superior. Some managers have much to do, using both sides of the A3 form. The A3 will never be as good as "Live management in practice, ", but should be the next best option.

Professionals often discuss the A3 as a scientific process. It is, however, purely a format to encourage good technical writing about scientific research and project management. Writing reports on A3 Hoshin projects presents a problem,, a company can use any world-class process of problem solving as a framework for scientific research.

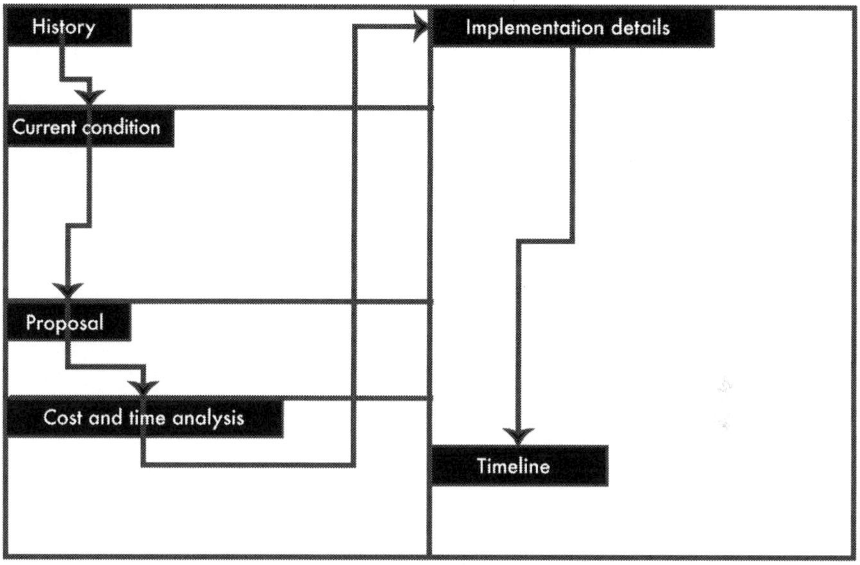

Figure 39 A3 Report Draft (adapted from SABEK II; SMALLEY, 2010)

If your company has approved these problem-solving processes as a standard, there is no need to switch to another process to use the A3. It is extremely flexible and as it spreads to other types of businesses and industries, such as health, can be easily adapted to any need.

The six A3s presented by Jackson (2006) are designed as an integrated set of writing and good promotion technique, more importantly, to support organizational learning about the processes of Hoshin Kanri and problem solving across the enterprise. Below is a brief description of each A3, approximately in the order they are used to manage the Hoshin process, along with the reference values for teaching purposes. The set of A3 to implement Hoshin Kanri are:

- Smart report (A3-i) and intended to build a consensus on the changing conditions of supply and demand before the A3-X. Used in the implementation phase of the Hoshin process.

- Matrix-X (A3-X), designed to unite severalA3-Ts together, explores the interdependence and relate all to the overall results. Used in the phase of "P - Plan" of Hoshin planning process.

- Team Plan (A3-T), proposes to carry out a strategic, tactical or operational Hoshin experiment, A3-Ts appear as "tactics" in A3-X. Used in the phase of "P - Plan" of the planning process Hoshin.

- Status report (A3-SR), applied monthly generates a quantitative assessment report of a PDCA research and is connected to the A3 and A3-Ts-Os. Used in the phase of "C - Check" of Hoshin planning process.

- Problem Report (A3-P), is a proposal of how a team can solve an immediate problem and not covered by the annual Hoshin. Used in the phase of "C - Check" of Hoshin planning process.

- Summary Status Report (SSR-A3) and a summary of the periodic report (based on A3-Rs) results of the A3-Ts and grouped in an A3-X. Used in the phase of "C - Check" of Hoshin planning process.

Hoshin Kanri for the Lean Business

To implement the first three steps and start with its Hoshin enterprise system, companies that operate then the Hoshin team will have to devote a minimum of six to eight weeks for these activities:

• Environmental analysis: This is research that the team must complete before conducting anything else. It is also a staff activity that should be performed throughout the year.

• Create a medium-term strategy. Hoshin team should spend a whole week building the medium-term strategy and discuss the intricacies of the process measures and financial results.

• Designing the annual Hoshin. This requires a week to fill the A3-X and, like most members of the team are Hoshin tactical team leaders, the establishment of these teams is easy. To Deploy Hoshin, the other teams will have about a week for deployment level, and another week to keep the Hoshin team; final meeting four weeks in total. The process may take longer when the organization has many plants in different states or countries and the involvement of suppliers will require even more time.

The Hoshin Kanri for the lean enterprise developed by Jackson (2006), explores the Lean A3 method to perform the Hoshin planning, as Hoshin management model (see Annex A).

4. ENTERPRISE DIAGNOSTICS METHOD PROPOSAL

We present in this chapter the Company Diagnostic Method for a continuous improvement tool, helping managers in decision-making in project implementation in synergy with employees and Hoshin more sustainable planning.

To get a quick sense of the magnitude of the proposal of this work, which applies the EDM in the business context, figure 40 shows the relevant details of the integrated strategic management company to diagnostic method and Hoshin Kanri model. This figure contains the interrelationship and the alignment of Enterprise Diagnostic Method with the operations strategy and Logistics upstream and Hoshin planning and strategic management downstream.

4.1 Introduction to the Enterprise Diagnostic Method

This research project is structured and aligned with the application of three management tools: the first is the diagnosis of research based upon the proven experience of the application of industrial benchmarking (VOSS, 1997), which reports the company's competitive position in several manage ment areas within the industrial sector of its operations and interest; the second is the VSM - Value Stream Mapping (VSM, Value Stream Mapping, translation Rother and Shook, 1998), which is a universally accepted language and in which, simply, there is at mapping of the value stream add value and an elimination waste, which means it is a new alternative to learn to see the opportunities; the third is the Capacity Management Survey, in which according to Hansen (2006) the enterprise can perceive the adoption and use of OEE as a measure for operational efficiency and requires an integrated management action on the shop floor focusing on maintenance , quality, process, groups of rapid tooling improvements, internal Logistics and results.

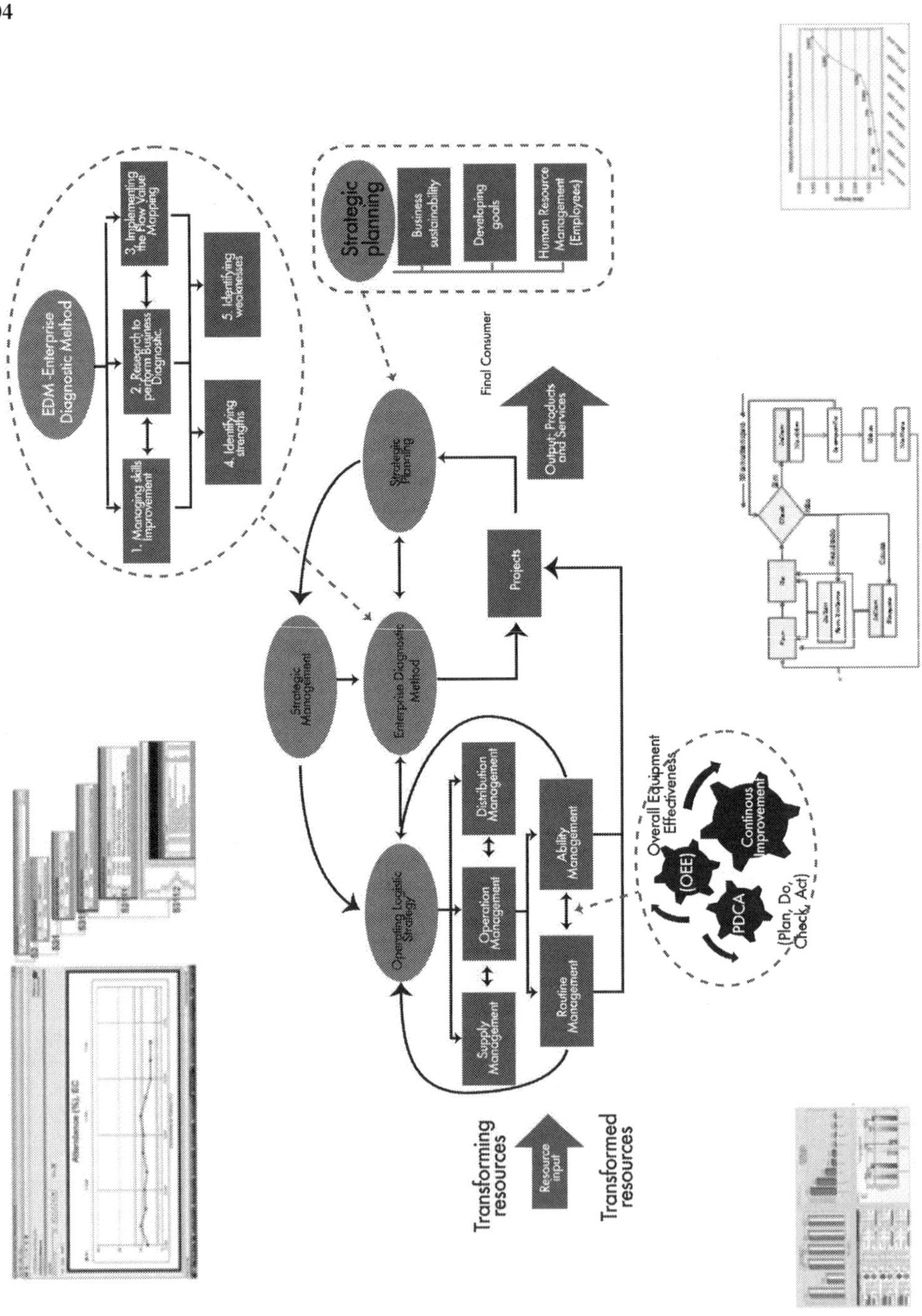

Figure 40 Integrated Strategic Management Model

Improving the strategic management and administrative processes within each company, begins with the preparation of a diagnosis that allows, based on their results, the drawing of an improvement plan to guide the necessary changes to improve management processes and achieve goals.

From the point of view of the proposal of the EDM, figure 41, the diagnosis of the company consists of three main areas: The application of Value Stream Mapping, the Management Research Capacity and diagnostic research firm; emphasizing that the self-assessment and survey evidence generate the identification of strengths and weaknesses, that in change translate into possible improvements and exist in the context of interviews with answers for the company's diagnostic research and develop management abilities for research that are answered in the workshop and, with the participation of the researcher. These sample forms are available in the technical report (CALADO, 2011).

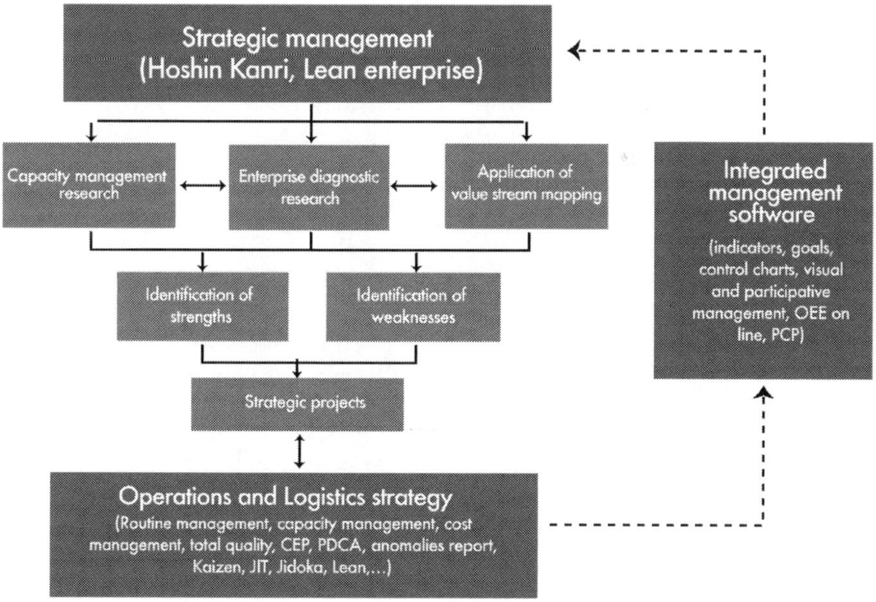

Figure 41 Enterprise Diagnostic Method - EDM

4.2 EDM definition and structure

According to Ferreira dictionary (2009), methodology is the scientific study of methods, which also means that is the art of guiding spirit in the search for truth. The method is an orderly way of doing things. It proceeds strategically and defines a technical process of calculation or experimentation effort to an end. The company is a commitment towards development of business, enterprise or corporation for the operation of an industry or service. A diagnosis is a label given by a specialist to a situation or current state, based on an assessment that is seen in the company, as the company is understood to be a corporation for the operation of an industry or trade.

According to these definitions it is understood that the Enterprise Diagnostic Method (EDM) literally refers to a method of study and qualification of the company's maturity level to be internalized by the organization, where it is observed, and assessing the results with mathematical method, in a participatory manner, by applying a set of methods and tools of science according to the Lean approach.

The material is geared especially for managers, management and technical staff of the companies that want to do a better Hoshin planning, strategic management, sustainability and find ways to contribute to achieve high level of maturity and/or turn the company into a world-class company with high performance through the application of best practices and Lean Enterprise. The application of diagnostic favors and helps the company's management model in the search results and continuous improvement.

EDM provides a continuous improvement cycle in the company as it is a system of analysis and strategic decision making that involves all the areas, levels and dimensions of the company, to improve the way you manage. It is coordinated by the direction of the company and conducted with the participation of employees and managers, consolidating the general knowledge of business processes to achieve the desired results by the leaders.

This process, that is illustrated in Figure 42 and described in the following steps, stimulates communication and integration of the professional group and develops a critical and proactive sense as the needs and maturity of the company, it is also a self-assessment on the best practices and performance with the free and open thought and practical and interdepartmental way. There is a developed shared responsibility among the participants of the company diagnosis.

The EDM promotes development and generates a broader knowledge about the processes of materials, administrative and information, practices and performances of the company, generate criticism, suggestions and participation in the strategic planning process directly or indirectly through exposure ideas, comments or justifications with every discussion assessing the issues of diagnosis.

In other words, with this method of diagnosing a participatory and interactive business participant's exercise the evaluators learn to listen, reflect on the different views, with the support of a researcher who assists in the construction of ideas or even with the clarification of the issues. We discuss technical terms unknown and as is understood and self-criticism, help in the initial construction of the cultural transformation of the company towards a world class and or competitive and sustainable organization.

4.2.1 The importance of diagnostics

The first step for improved strategic management in the company is to gather relevant information to understand company´s current situation. It is important to consider that while those responsible of the leadership of continuous improvement process involving the day to day operations of the company do not necessarily have timely and reliable information on the reality of the organization.

The EDM process involves collecting and systematizing information related to various aspects of business management which are not always known by all and usually a lot of information is not expressed and used in decision-making.

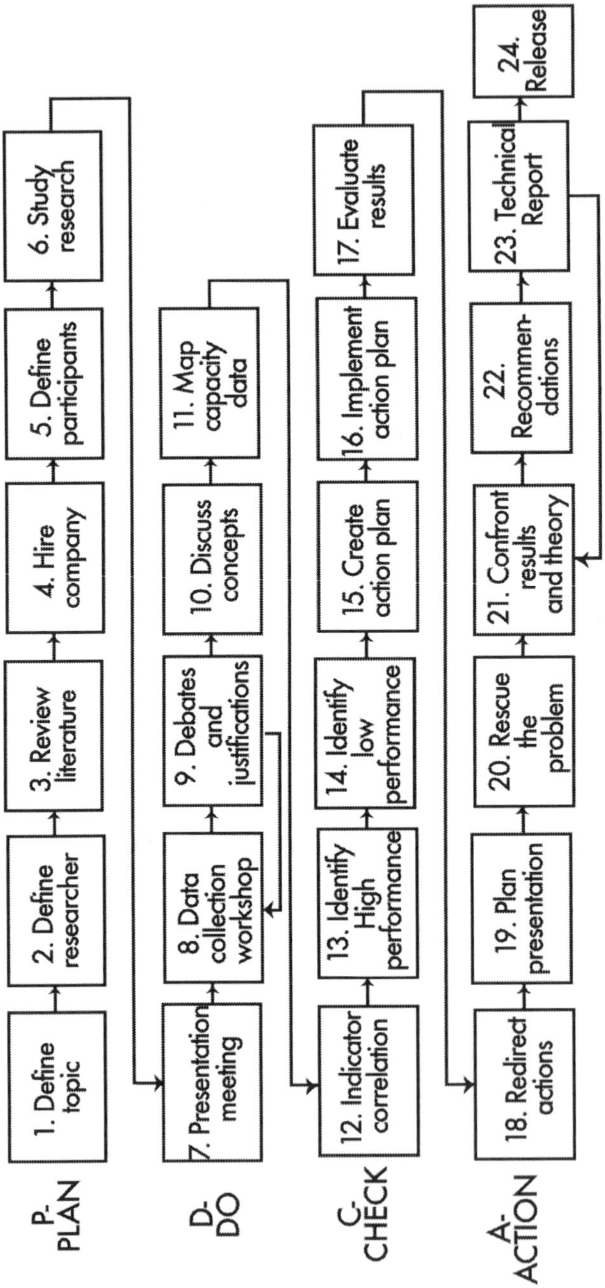

Figure 42 Enterprise Diagnostic Method Structure

Secondly, the continuous planning considers the design, evaluation and implementation of a strategic plan based on the results of the company's diagnostic process in an annual period or compatible with the company's need to be set. The formulation of the strategic plan involves making decisions regarding business development opportunities, aiming at a sustainable business and especially profitable.

As a continuous improvement of decision making, Hoshin planning and strategic management sees the completion of a review by an external body, and that professional or researcher has the role of moderation, reconciliation, integration and leverage teamwork, based on workshops and the approach of action research. Although this is a strictly voluntary activity on the part of the researcher, it is important to consider that an external character assessment ensures, from the application of standards and methods developed and tested by the researcher.

4.3 EDM stages

According to the EDM structure, this method is composed of twenty-four phases, the preparation and the development of theoretical knowledge and simultaneously the search for a participatory form of decision-making and Hoshin planning and strategic management support is used as the action research approach to develop the twenty-four stages of the EDM (table 7).

Table 7 Enterprise Diagnostic Method Steps and Descriptions

Stage	PDCA	EDM Main Activity Description
1	P – *Plan*	Define the topic and the preliminary research proposal: Apply the Enterprise Diagnostic Method.
2		Set up the research team.
3		Proceed to review relevant literature
4		Initial contact with the company interested in the Company's diagnosis.
5		Identify research participants
6		Study the feasibility of applying the action research method.
7	D – *Do*	Participants meet to discuss about the problems of the group or organization under investigation and action possibilities.
8		Based on a diagnostic assumption, collected up data for the preparation of diagnosis, through questionnaires, interviews, observations, workshops (participatory meetings) and other procedures.
9		The researcher promotes debate on the assigned score for each question based on the integration and collaboration of the participants.
10		The researcher promotes discussions based on modern concepts of production, Lean Enterprise and *Hoshin* Planning (*Hoshin Kanri*).
11		Intensify data collection, mapping the value stream and assesses the capacity management.
12	C - *Check*	Elaborate calculations after analysis of diagnosis with Grey correlation analysis method - selection and evaluation indicators.
13		Select the high performance recommended issues according to the Grey correlation analysis method.
14		Select the underperforming recommended issues according to Grey correlation analysis method.
15		Elaborate the action plan, considering the actions to be implemented, those responsible for implementation and deadlines.
16		Implement the reporting actions with best practice recommendations
17		Evaluate the result of each of the research questions in practice and the development of theoretical knowledge.
18	A - Action	Redirect actions, if deemed appropriate.
19		Plan to present the research results.
20		Go back to the problem that prompted the investigation
21		Confront it with the results and theories obtained supported from the research
22		Execute a list of recommendations
23		The research report is elaborated and made the presentation of the research diagnostic project.
24		Disclose research results.

4.3.1 EDM planning steps (Plan – PDCA)

Considering the PDCA problem-solving method, the EDM planning occurs during the first six steps; that don't have to be executed necessarily in that order; described in detail:

Step 1 - Set up the theme and the preliminary research proposal: Applies the Enterprise Diagnostic Method in companies the researcher's contact network, at first it is not a consultancy but research for development of the strategic management of the companies.

This work is the result of necessity and empirical research conducted in recent years. It is indeed an energy drain for professionals and a waste of money when the result of the implementation of techniques and modern methods of production do not coincide and have no integration and alignment with business objectives.

The main idea is to identify best practices and opportunities for improvement for decision making, according to the needs of each company, therefore, was taken as a premise use Benchmarking, Capacity Management and mapping the value stream to identify and interact with the companies surveyed.

Step 2 - Set up the research team: The researcher is the facilitator and moderator responsible for developing and coordinating the implementation of the EDM. Applies to an interdepartmental team, i.e., the workshop participants are business professionals surveyed with total control, knowledge of operational and administrative processes of the business and will be appointed by the directors and or managers.

Step 3 - Relevant literature will be reviewed: The researcher was initially based on literature on benchmarking, mapping the value stream and OEE, and extends the studies, aligning them with the research on Hoshin Planning and Lean Enterprise. Figure 41 depicts the main research themes. In the absence of a researcher hold that a company project manager, after being trained, drive EDM.

Step 4 - Initial contact with the company interested in the Enterprise Diagnosis: An explanation of the research proposal to company management is created. It is clear, in this first meeting, which provides the research project, that the company will provide diagnostic aid subsidies in investment decisions, strategic information and opportunities for process improvements, business people and with the participation of the very people the company. This provides arguments of the advantages of identifying best practices to generate continuous improvement projects in which you can improve business performance if there is Hoshin planning.

The disadvantage identified and discussed is the risk of the confidentiality of company information, but verbally an informal or formal confidentiality agreement is established and it is also correct to submit the final report in the first instance to the responsible director.

The researcher will act as a moderator and personally lead the workshops (research meetings and continuous improvement), and at times be an agent clarifying the questions from participants about concepts and, whenever possible. Use the Brainstorming and other techniques of moderation to conduct continuous improvement leverage and improve creativity on the participant side. A company project manager can lead the workshop in case they are trained in EDM.

With the aim of greater openness and confidence in direction on the proposed research project, a summary of work history research is presented, showing their experience in business, research and academic. There are cases that the company's director delegates a trusted manager to coordinate workshops, they will be responsible for the internal organization of workshops, preparing internal agenda, invites participants, booking room and projector in the company and gives due support to the researcher.

Step 5 - Identify the research participants: Shape up the team under the guidance of the responsible director or manager, also called facilitator or coordinator of the company diagnostic research, the work team with technicians, operators, leaders, managers with extensive knowledge the areas and the company. Whenever possible the participation of at least seven surveyed areas is recommended: Quality, Production, Logistics, Organization and Culture, New product performance, management Innovation and Environment, Health and Safety.

At this point, the researcher makes the quest for alternative operators to include in the project to have a voice on the factory floor with the diagnosis and in the future, participating in strategic planning. This action alone generates a behavioral change from conventional and centralizing management in addition to already show everyone the need for participation of all levels in participatory planning.

This team tends to be part of the committee that prepares the Hoshin planning seeking synergy and business sustainability and collaborates directly with the strategic management and business results.

Step 6 - Study the feasibility of application of the action research method: Given the experience of the application of action research in the implementation of techniques and methods of Lean Production in environments with conventional production system, the researcher chooses the moderation techniques, questionnaires, interviews and action research.

4.3.2 EDM execution stages (Do – PDCA)

Continuing the comparison with the PDCA, step seven and eleven step run up the workshops of diagnosis and are detailed below:

Step 7 - Participants meet to discuss about the group's problems, the organization under investigation and action possibilities. In this first contact between researchers and the participants in the meeting address the following issues:

- The management representative introduces the researcher to all selected participants and department representatives. At the beginning, the director expresses support for the research project by the need to have consistent information for decision making, achieve goals and results of the company in an integrated, aligned, participatory and sustainable.

- A dynamic presentation follows in which the researcher gets to know the participants, their experience in the company and its area of responsibility and work, among other information.

- The researcher with the intention of leveling of information between the different levels of selected participants, provides a brief explanation in the form of lecture on the competitiveness of enterprises, the difference between a conventional company and a modern enterprise, explains about the history of the Toyota Production System and Lean enterprise, the need for customer value and reduction of waste to increase the assets of an integrated, participatory and sustainable business.

- The researcher explains to the participants the purpose of the EDM workshop, as agreed with company's management, as explained in Figure 41.

- Explain to the participants that this work is intended to support decision-making and will be critical to the Hoshin planning, also explains that such identification of weaknesses, judged by the participants, will be used to leverage and create a list of recommendations improvement projects or maintenance.

- Explain that the method starts with the participation of employees with direct and indirect positions in operations and logistics strategies, which includes representatives of the administration and service areas, if applicable (1), with the participation of sponsors in the case employees of the management and direction decision maker and assets (2), applies the diagnosis as a self-assessment process (3), the researcher prepares a final report with analysis and list of recommendations (potential projects) and identification of level of maturity of the company according to world-class company approach (4), the implementation of an annual strategic planning is proposed and if necessary with six-monthly review (5) Finally, explain about the monitoring and control of results through strategic management (6)

- Explain that actual results are achieved with the implementation of the six stages mentioned and implementing a systematic way, combined with problem-solving methodologies as PDCA and A3 report, if it is part of our routine, because by applying the Diagnosis, strategic planning is drawn with synergy and the business management is focused on the bottom line with sustainable business vision. It is expected that from the first diagnostic meetings the culture of the members will begin to change, undergoing some changes given that the reviews discussed in the meetings through their participation, increases the degree of knowledge about the Lean concepts and self-criticism of participants promoting hassle and discomfort for the conditions of waste and the trend is that everyone will start spotting missed opportunities to add more value to the customer.

- Explain that this research project includes the first four steps free of charge and voluntary and the two other stages of Hoshin planning and strategic management are suggested and guided if there is interest from the business.

- The researcher explains success stories of using Hoshin kanri and the possibility of having a track of all actions and projects in an integrated way in the 6th step to better management, visibility and management with monitoring of financial results.

- After presentations and clarification regarding the EAW, the researcher presents a guidance schedule already agreed on with the direction of the company that contains the following topics:

1 – Presentation of the proposal and approval of the project by the EDM direction.

2 – First workshop day, meeting with participants and direction to understand the project scope.

3 – Second workshop day, responds and discusses the issues for the diagnosis of Com-

pany: Organization and Culture (11 questions); Production cycle time (six questions).

4 – Third workshop day, working on the topics of: Quality (10 issues); Plant and equipment (10 questions).

5 – Fourth workshop day, working on the topics of: Practices and engineering operations (6 issues); Innovation (8 issues); Process and product development (5 questions).

6 – Fifth workshop day, working on the topics of: Evaluation and performance in the company (8 issues); Investments (4 issues); Environment, Health and Safety (12 questions).

7 – Sixth meeting, plans to research the capacity management (21 questions).

8 – Seventh meeting plans to value stream mapping execution along with the participants'. Attempt to show opportunities on the shop floor or field.

9 – Eighth meeting is reserved to present the EDM to the participants and management guests. The latter meeting indulges towards a final report of the EDM prepared by the researcher.

The average workshops last for three hours with a 15 minute interval with the total duration of 25 to 30 hours of work in the company for each participant. Considering the involvement of seven employees, we have on average 210 hours and added another 40 hours of researcher's work; the overall total is estimated to 250 hours for each research project and each company. This information varies with each company's reality and consensus with management.

The average search period is two to three months, because the meetings are usually scheduled weekly and according to the availability of the agenda of employees, as this agenda tends to favor the integration of the researcher to the group of participants, leaving them to comfortably and confidently construct critical opinions on the issues that will be discussed about the business.

The staff invited to every meeting is always the same group of individuals, it is a chance to meet the particularities of other areas, give opinions on the opportunities for improvement and put their point of view on each issue assessed relevant to different areas.

The purpose is for the moderator to receive open and verbal feedback from participants about the research and methodology issues in general. Through researcher moderation participants interact on the reality between departments and how the workshops take place at each meeting, participants are open to discussing their problems with colleagues and can now take proactive actions to improve management the routine day-to-day.

Step 8 - Based on a diagnostic assumption, collect data for the preparation of the final report of diagnosis, through responses to questionnaires, interviews, observations, workshops (participatory meetings) or seminars and other procedures.

First the researcher explains the issue and the meaning for the participants to attach notes 1 to 5. At this stage, participants can hear and follow the explanation for the slide projector for each question of the questionnaire.

Second, the participants ask questions to the researcher on the issue and the terms that cause doubt and or unawareness. When it comes to a company culture and conventional production system is common to the researcher to explain the concepts related to the questions. If a participant has doubts the explanation of the issue then its main concepts are explained by the researcher, favoring the learning process and knowledge of all on each question.

Third, participants are asked questions in written form, one per participant and then describe a particular issue in standardized form.

Fourth, once the issue is discussed, participants describe the approach and its justification referring to the note, that way, participants have the opportunity to suggest improvements and or make criticisms of the issues discussed. All the marks awarded are followed by justifications.

Step 9 - The researcher promotes debate on the assigned score for each question based on the integration and collaboration of the participants: Participants discuss the notes and expose their views on the issue, which also happens to be a more detailed discussion of the issue in the company, with the discussion the moderator promotes continuous improvement of vision. This type of discussion leads the group to a self-analysis on where and how this can be improved, all come to an understanding of the good and bad parts in every issue.

10th stage - Researcher promotes discussions based on modern concepts of operations and Logistics, Lean enterprise and Hoshin planning. Participants also collaborate with their expertise. It promotes the understanding on the need to implement problem solving methods and strategic management to increase the company's assets.

11th stage - Intensify data collection, value stream mapping and assess the capacity management. Undertake the future value stream mapping along with the participants according to the VSM methodology. Look at the current mapping in order to confirm the waste already mentioned from the diagnostic meetings. Also proceed with the response capacity management questionnaire with dynamic interview and moderation and promoting discussion of each question and answer. The researcher draws up a list of recommendations based on the justifications and marks awarded to improve the company's performance, and the information is included in the final report.

4.3.3 EDM verification stages (Check – PDCA)

This is the penultimate stage of comparison with the PDCA, from the twelfth step, up to the seventeenth stage, the researcher will transform the data into information and are described below:

12th stage - the calculations are done after the diagnostic analysis with Grey correlation analysis method for selection and evaluation indicators. This is when an analysis method, the selection, and evaluation is selected forth indicators that will be used for understanding, planning and preparation of the final report to make a decision.

- Execute company rank according to the analogy to the boxing match according to the benchmarking to know the level of maturity of the company.

- Assess the current situation to achieve better performance against the marks awarded to the practices and performance (Performance).

One can get historical information on the industrial benchmarking database in IEL/SC and compare the researched company and the average of the worldwide industry leaders. Analyze the company's position in relation to quartiles in practice and performance in the sector. The quartile is a variable value in question allocating or dividing the amount of percentage elements above or below.

We adopt the box plot tool method, which defines a statistical measure called the separatrix first quartile (Q1) that establishes a cutting line for the data set. The first quartile (Q1) separates the bottom 25% of the ordered values higher than 75%, more particularly at least 25% of the ordered values are smaller than or equal to Q1, and at least 75% of the values are greater than or equal to Q1. The calculation is done with the help of Minitab software, and the worst indicators according to the first quartile are defined for each company.

The second quartile (Q2) is the same as the medium which separates the lower 50% of the ordered values of the upper 50%. In the third quartile (Q3) there is a separation of the lower 75% of the ordered values of the 25% higher. Therefore is possible to calculate the result of the third quartile with the Minitab software and identify the cutting line for 75%, identifying the best indicators of this set of data, after selecting and evaluating method Grey correlation analysis. So it is possible to know the company's maturity level by applying the boxplot in a set of business indicators. Later you can evaluate and see the new level of maturity through a diagnosis of the company and use the boxplot.

While the comparison between companies in the same segment may be used, EDM does not focus in comparing companies, but the identification of the classification (ranking), the selection and evaluation of the indicators, i.e. a cut line identification the strengths and weaknesses using the first quartile and third quartile of the boxplot tool.

- Through the trend graph display is the company's position in relation to the contents of practice and performance

- Then a comparison is made with the average of the worldwide industry leaders.

- To identify Strong and Weak Practices indicators, the grey system is used.

- To identify the Strong and Weak Performance indicators, use the grey system.

13th stage - Select questions and commendations that had a high performance or that got higher grades or marks by the evaluators. Elaborate diagnosis after analyzing the data for information and maintenance of the organization, according to the Grey correlation analysis method.

14th stage - The underperforming questions, recommendations are selected according to Grey correlation analysis method. Elaborate diagnosis after data analysis, improvement opportunities for information and organization of continuous improvement planning.

15th step - Elaborate the action plan, considering the actions to be implemented and those responsible for implementation and deadlines. This is a step in the industrial benchmarking application method in which their facilitators coordinate a planning session with the benchmarking team company, assisting in the creation of this action plans to explore opportunities for improvement identified in comparison with leading companies.

This research did not carry out the action plan, as it is understood that this is an activity that belongs to strategic planning.

16th stage - Implement the reporting actions with recommendations of best practices, based on both research and VSM.

17th stage - Evaluate the result of each of the research questions in practice and development of theoretical knowledge.

4.3.4 EDM standardization steps (Action – PDCA)

In this last phase of the comparison with the PDCA, from step eighteen to step twenty-four the results are presented and this prepares the environment for strategic planning, as details below:

18th stage - Redirect the actions if deemed appropriate.

19th stage – Plan the search results presentation.

20th stage Discuss the problem that prompted the investigation.

21th step - Confront with the result(s) the Theory(s) that supported the research.

22th step - Execute the list of recommendations, direct to the application of best practices in a personalized manner by business need.

23th stage – Elaborate the final research report and make presentation of research diagnostic project. Discuss previously with management.

24th stage – Disclose search results

During the interviews hunches and guesses may occur with possible involvement of emotional or political nature. As compared to other companies this is not the main approach of diagnosis, diagnostic research does not follow the same methodology of industrial benchmarking. As the questionnaire is completed during the interviews, it is possible to gain insight into the answers. It is noteworthy that among the interviewees there is inhibition, criticism, resistance to answer, and this generates a good level of interpretation of the questions because it is a special condition in a participatory manner by employees, directors of companies and the researcher.

4.4 Enterprise diagnostic method basic principles

There are differences between knowledge, skill and attitude among employees at all levels, in particular, in the matter of employees knowledge participation and satisfaction it is necessary to carry out improvements that enhance their quality of life and bring results for the company is emphasized you must adopt are some basic principles:

a) The management is based on a thorough understanding of the needs and expectations of employees and management of the company.

b) The vision and the management strategy consolidate the contribution and how the synergy will perpetuate sustainability in business and meet the needs and expectations of stakeholders.

c) The members of the company are intended to contribute to the improvement of production and administrative processes.

d) The continuous improvement and learning processes in the company based on Lean performance standards and practices must be monitored through the indicators within Hoshin planning.

e) The results are known after diagnosis, analyzed and reported externally by management and responsibility is assumed.

4.5 Integrated EDM into strategic management

According to the proposal of the procedural model of integrated implementation company to diagnosis and planning (see Figures 40 and 43) six areas can be distinguished in the company's management. These areas will guide the strategic management process of building from the EDM. It begins with the guidance of the direct and indirect employees in active operations and Logistics (1), guidance and direction to the management decision maker and assets (2), a self-assessment process (3) recommendations list formulation and level of maturity of the company in accordance with the approach of the company Diagnostic method (4) implementation of an annual Hoshin Planning (5), monitoring and tracking results through strategic management (6).

Guiding employees in operations and logistics - Output: This area refers to the way the company manages the routine day-to-day, manages the capacity of the equipment, its employees, their expectations and their levels of satisfaction. Attends how the company promotes and organizes to meet the markets with their Lean principles, with Just-in-time and Jidoka approach, adding customer value and continually reducing waste.

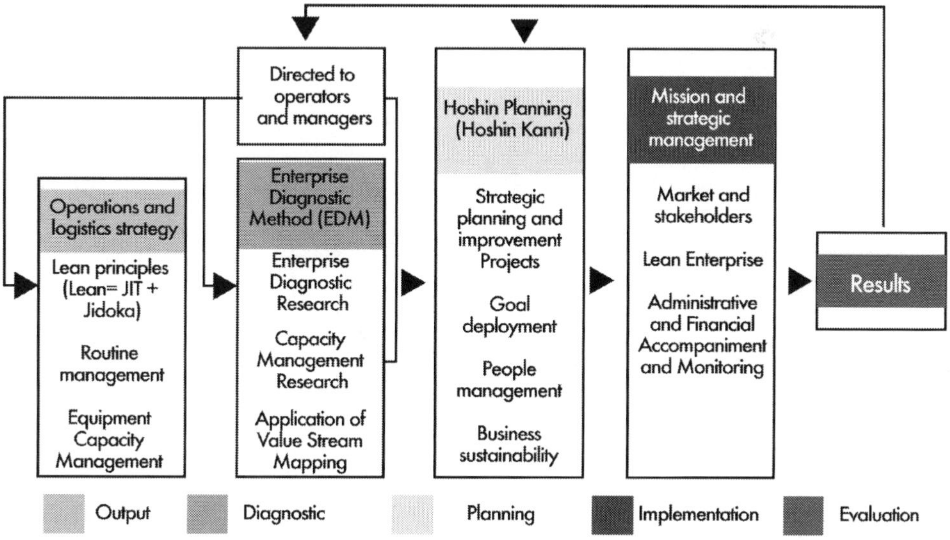

Figure 43 Model of integrated strategic management of EDM and Hoshin Kanri

Companies seek to make full use of workers' skills, materials and information, which is why the JIT and Jidoka are adopted due to the fact that any problem in the process or in the supply chain can affect production. The result of the JIT approach is the reduction of the lots, inventory and lead time and Jidoka approach changes to the culture of the operators, as they are always attentive and aiming to eliminate waste, only doing what is necessary and with quality.

Guidance for the management and management decision makers and assets - Diagnosis: this area looks at how the management guides and leads through information systems and procedures for Hoshin's management planning and the results integrated into corporative goals. At the same time, it deals with the way the direction performs management strategies with a view to the satisfaction of employees and customers, solving problems and how to reach the goals and evolves towards world-class company and sustainable.

Process of self-evaluation - Diagnosis: This area takes into account the existence of the professional skills profiles for the diagnosis. It also considers the existence and operation of the existing organizational systems and the commitment of the training level and or improvement work team, whether they are professionals, managers or not, in order to push them on the company's direction and representative areas. .

The diagnostic method of research studies the capacity in competitive companies based on seven critical areas: Total quality, Lean Production, Logistics, Organization and Culture, New product performance, management Innovation and Environment, Health and Safety. The area of New product performance and Innovation management in the research is evaluated only in companies that develop products in the evaluated operation, but for the purpose of the MOU was chosen for use in full and evaluate all the aforementioned areas.

Final report with recommendation and level of maturity - Diagnosis: refers to the preparation of a technical report that is delivered to company management. This paper details the application and is structured in seven topics with content: the first chapter covers the objectives of the EDM and a brief account of the company scenario, according to the Statistical Yearbook of Production Development Secretariat. In the second topic, describes the steps of the EDM, the third topic describes the application of management research and the performance of the overall efficiency of equipment (OEE), it will also present the application of value stream mapping in the current process and research on the company diagnosis.

The fourth topic shows the company's ranking according to the benchmarking methodology and, in more detail to assess the strengths and weaknesses of the company ordered according to correlation of gray correlation analysis system.

The fifth topic includes a list of improvement actions based on justifications of the evaluations of participants and a conclusion of the researcher and facilitator to direct Lean improvement projects, the goal statements, management's actions upon human resources in a sustainable way that considers the strategic objectives.

In the sixth area the researcher presents a brief bibliographic reference on relevant matters and, finally, the topic seven attaches forms used and operational and or administrative information processes.

4.5.1 Hoshin planning – planning phase

This area is considered the systematic development of the company's processes that have been researched. That will be detailed in the Hoshin planning work, but when it comes to strategic planning, according to Certo and Peter (2010), it is important to explain about the corporate strategic management and business:

- Strategic management evolved from financial planning, it is embodied in the budget for the long-term for strategic planning.

- From the strategic decision-making model, strategic planning has evolved into a broader model, which is based on a series of interrelated steps, so that a change at some stage of the process can affect others.

- Strategic management is being viewed as a systematic approach to change management, comprising: positioning the company through the strategy and planning capabilities; strategic response in real time by administering issues; and systematic management of the resistance during the implementation of the strategy. And after the formulation of the main goal the process to formulate the Lean principles and objectives should begin bearing in mind the organization's strategy, which takes place on three levels: Enterprise level, Business unit level and Functional level.

- The Enterprise or corporate strategy, depends on business conditions (size, industry, etc. structure), and can be considered as the most important in the business context, given the scope of the strategic decisions that it comprise, among which it determines the direction and objectives within the organization, with impacts in all its business units and functional areas.

- The corporate strategy is presented as comprehensive decisions, such as setting the corporation's boundaries (decisions on the scope), the direction of corporate decisions about the relationships between the business units and the establishment of the methods that define the degree and the form of corporate diversification.

- The business strategy, despite its concentrated development in senior management, must be known by all employees of the organization, which should act in a participatory way in its implementation.

- The set of decisions that are under the business strategy include influences, the principles, objectives, mission and direction of the organization, as well as impact the sustainability of the business, and significantly on the profitability, competitiveness and degree of satisfaction of its various stakeholders.

As an example and to get a better idea of the number of steps of the activities that are developed during the strategic planning levels, see Figures 35 and 36.

To implement the strategic planning in Hoshin style, according to Jackson (2006), teams must be organized in three levels and unfold the top-down targets, once they organized the Hoshin teams perform:

• Analysis of diagnosed environment. This is subsequently evaluated on the EDM and it is suggested that this activity should be accomplished throughout the year.

• Create a medium-term strategy. The Hoshin team should spend a whole week, building the medium-term strategy and discussing the methods of process measures and financial results.

• Designing the annual Hoshin. This requires a week to fill the A3 model and, like most members of the team are Hoshin tactical team leaders, teams should be driving the efforts. The process can take longer if the organization has many production plants indifferent places and wishes to engage its suppliers.

When planning strategies are based on business principles, principles apply to all areas of the company. Table 8 illustrates some basic tools and methods used in the implementation of a production system with an approach to Lean Production, the same is presented in detail, indicating references in chapter two.

When it comes to processors, such as the surveyed in this study, the objective of increasing its assets (make money) becomes a global goal taking into account Speed, Cost and Innovation. The goal of revenue, operating profit and cash flow will be achieved with the competitive performance guidelines given to the Company and shareholders, for customers and for employees. For the company and its shareholders interests: revenue, return on sales; market share and margins. For customers interested: quality in manufacturing and use; warranty costs; quality award. For employees: safety, ergonomics; good working conditions; training opportunity and development; employment stability. For the community leaders interested on environmental issues involvement in communities and stable employment.

Dennis (2006) illustrates and describes the performance objectives are achieved with the strategic and aligned deployment to the factory floor with the standardization work (Figure 44).

Table 8 Lean Principles, Methods and Basic tools

Lean principles (categories)	Methods	Tools / techniques
Produce demand	Pull System	Analysis ABC / XYZ; JIT - JIS; kanban; milk run; supermarket; VMI
	Takt time	One piece flow; Leveling of production; Quick change; VSM
	Supply management	Marketing supplies and selection of suppliers; Integration with suppliers; supplier management
Robust product	Product variants management;	Product order release
	Management phase-in and phase-out	
	Project management	Planning Roadmap; Resource planning and time; Risk management
	Product development process	Product development process manual
Employee motivation	Working in teams	Climate research; Dialogue with the team; responsibility matrix
	Qualify and develop employees	Specific and custom tools
	Performance orientation	Specific and custom tools
Avoid defects	Defect prevention	FEMEA; *poka-yoke*; CEP
	Fault detection and planning tests	Line stop concept; Product audit; self-Control; Reliability test
	Analysis and correction of defects	Registration of production errors; Defect analysis field; PDCA; Analysis of returns
	6 Sigma	Boxplot; Histogram
Avoid waste	Standardized work	The post organization - 5S; Working standard, TPM; OEE
	Visual management	Management of the indicators; Identifications and demarcations
	Continuous improvement process	Troubleshooting techniques; Workshops improvements

Dennis also suggests using the A3 report for the planning of activities, but Campos (2004c) describes the deployment of goals can also be done using the tree diagram and a diagram matrix, a way to illustrate the split in the industrial area is through of Figure 45, in which an example of prioritization matrix measures and execution of projects is shown, in a simple and detailed manner. (Figure 46 and 47).

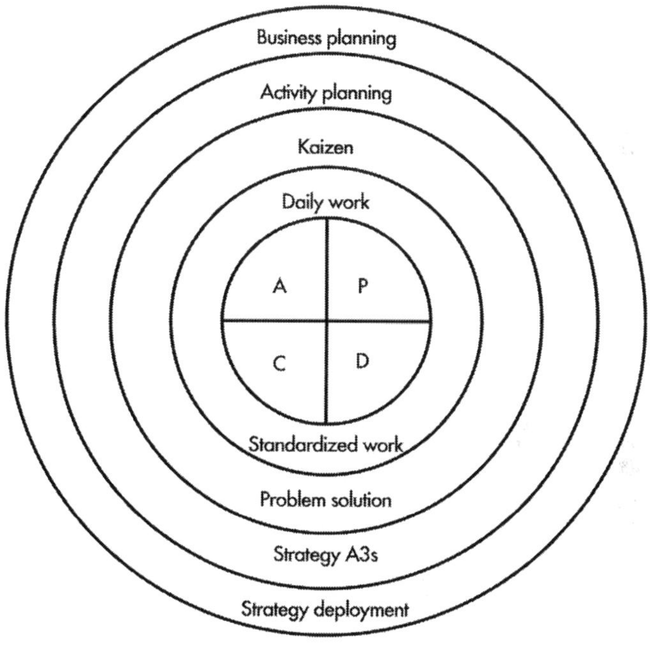

Figure 44 PDCA in Lean organizations (DENNIS, 2006)

The Tree Folding Goals for Industrial Areas in Figure 45, was a way to manage the goals from the 2nd level of the company (industrial direction), in which all departments (managers) and sections (supervisors) joined the visual management model all followed the same Prioritization Matrix form of model measures and Project Execution. The use of this tool has brought employees and managers the visibility of goals and guidelines, which strengthened the weekly feedback meetings of managers and supervisors and feedback at monthly meetings with the industrial director

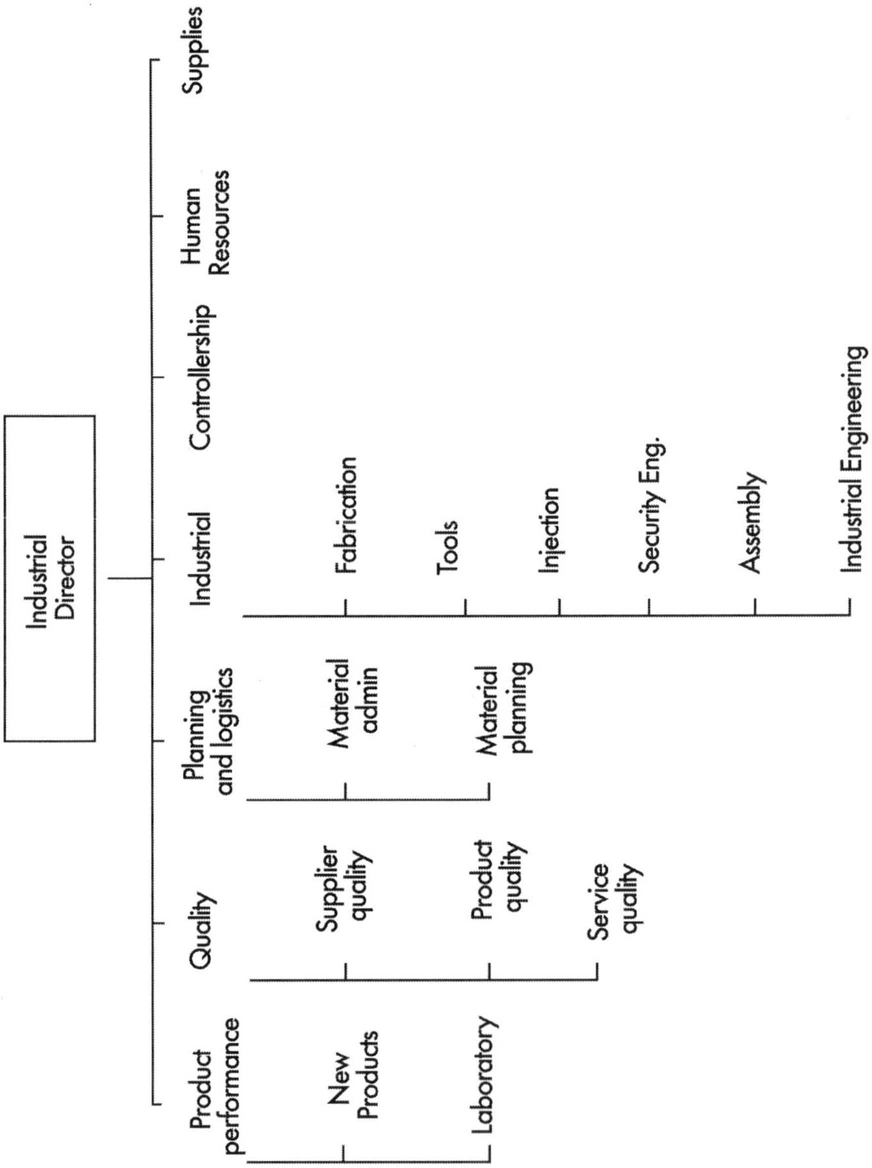

Figure 45 Outline Goals Folding Tree for Industrial Area

Guidelines for 2006 - Industrial Department

Objectives			Delivery		Productiv. MO	Quality		Costs				Safety	Environment
Projects/Goals			Production capacity from 2,200 prod / day and 2,500 prod / day	Improvement of the factory weekly OTDF for a minimum of 95%.	Ensure minimum Productivity MOD of R$ 63.5 / h (1.5% un 2005)	Raise the average annual value of QKZ to 97.5%	Raise the average annual value of FPY to 95%	Ensure maximum cost value in the Department (OD) of R $ 46,760,598.00	Ensure a maximum value of WIP (Deposit 108) of R $ 2,000,000.00 in the Industrial Department	Assist in MCR projects seeking cost reduction of R $ 16,500,000.00	Reduce Scrap and Rework index of Industrial Department for a maximum amount o f R $ 1,800,000.00.	Reduce Accident Frequency Rate of 11.5 to 8 (no accidents x 106 / hours worked)	Implement 100% of the projects outlined in PGA
Nº	Main projects to achieve goals	Project Description	15	10	10	5	15	10	5	5	5	10	10
1	Plan Factory	Improving Quality indicators, WIP and productivity through the implementation of new forms of manufacturing and internal logistics management	●	●	●	○	●	○	●	△	●	●	○
2	Production Firm Program	Ensure the supply of parts with at least two working days prior to the production program. The supply includes quantity and quality	●	●	●	○	●	○	●	⊗	●	○	○
3	LEAN PROGRAM	Continue implementation of the HPS program (now called enterprise Lean) including all its tools aimed at reducing the various types of waste material in accordance with the philosophy of Lean Manufacturing	●	○	●	○	○	●	●	△	○	○	○
4	MPT	Total Productive Maintenance. Is reflected in the improvement of OEE and reduction of costs and accident rates	●	○	△	○	○	●	○	⊗	○	○	●
5	Reduced Rework	Reduce the need for rework cabinets and doors foamed by damage reduction throughout the process	○	○	△	○	●	○	○	⊗	●	○	○
6	Reduced Standard Time	Improvement of working methods and accuracy of data to achieve a reduction in the standard time of manufacture of the products.	●	○	●	△	△	○	△	△	△	○	△
7	Operating standardization	Deploy the MO Certification (Fit's + Training + Audit + System Maintenance) in the Manufacturing sector	△	△	○	●	●	⊗	△	⊗	○	●	○
Cont of Method	INDICATORS (Control Item)		Production Capacity	Weekly Delivery	MO Productivity	QKZ	FPY	QKDO ControlZ	WIP	MCR	Scrap	Frequency Rate	PGA schedule
	2005 History			77%	xxx	95.4%	93.5%	R$ 39,305,242.00	TBrl 6000	TBrl12 mil	TBrl 4400	TBrl 4401	TBrl 4402
	Measuring frequency		Monthly	Weekly	Monthly	Monthly	Monthly	Monthly	Monthly	Monthly	Monthly	Monthly	Monthly

Figure 46 Prioritization Matrix measures and Execution of Projects: Projects and Indicators

Guidelines for 2006 - Industrial Department

Objectives			Project Management					
Projects/Goals			Control Method			Responsible team	2006 Schedule: J F M A M J J A S O N D	2005 History
			Priority	INDICATORS (Verification Item - Formula)	Project Goal			
Nº	Main projects to achieve goals	Project Description	100				Planned	
1	Plan Factory	Improving Quality indicators, WIP and productivity through the implementation of new forms of manufacturing and internal logistics management	430	MOD	**Reduction to 425 MOD p / 2200**	Industrial Team leader	2007	**454 p/ 2200**
2	Production Firm Program	Ensure the supply of parts with at least two working days prior to the production program. The supply includes quantity and quality	405	Production losses Index	**Reduce 90% of production losses due to lack of parts (internal and external)**	Logistics Team leader		**12% of losses due to lack of parts**
	LEAN PROGRAM	Continue implementation of the HPS program (now called enterprise Lean) including all its tools aimed at reducing the various types of waste material in accordance with the philosophy of Lean Manufacturing	370	LEAN Score	**Average of 3.5 in audits**	Industrial Team leader		**average of 2.6**
4	MPT	Total Productive Maintenance. Is reflected in the Improvement of OEE and reduction of costs and accident rates	335	Loss of production for maintenance	**Maximum loss of 1.2% of planned**	Manufacturing Team leader		**1.7%**
5	Reduced Rework	Reduce the need for rework cabinets and doors foamed by damage reduction throughout the process	305	Reprocessing Doors index Reprocessing cabinets index	**Reduce rework to 1.5%**	Production Team leader		**2.17% Cabinets 2.4% Doors**
6	Reduced Standard Time	Improvement of working methods and accuracy of data to achieve a reduction in the standard time of manufacture of the products.	260	Time Percent Reduction	**Reduction by 5%.**	Industrial Team leader		
7	Operating standardization	Deploy the MO Certification (Fit's + Training + Audit + System Maintenance) in the Manufacturing sector	255	Multifunctionality Matrix	**Staff trained 100%**	Fabrication Team Leader		**30%**

Control Method	Legend			
	Weight	Project influence in Goal	symbol	
INDICATORS (Control Item)	-	No Influence	⊗	0
	1	Weak	△	1
History 2005	2	Moderate	○	2
Measuring frequency	5	Strong	●	5

X Company

Figure 47 Prioritization Matrix measures and Execution of Projects: Projects and Indicators (continued from Figure 46)

4.5.1 Integrated Strategic Management – Implementation and Evaluation Phase

Strategic management means the operation of a system and measurement procedures and evaluation of organizational performance. Not detailed in this job description of the EAW. It occurs in all levels and areas in the short, medium or long term, depending on the mission, Lean principles of the company, Strategic Objectives and Company Goals.

Strategic management is summarized in the systematic monitoring, analysis of management results periodically review of administrative and financial performance, which guide all decisions at the corporate level macro or corporate, business unit and functional, integrated and transparent manner to achieve better organizational performance and step by step, closer to being the planned operational excellence.

One of the current forms of strategic management is the development of a Hoshin planning, which requires development and filling A3 report templates with formalized teams in corporate levels, business units and functions, previously set by superiors. The Hoshin team members are often leaders and team managers and have a deployment profile and driving attitudes. Strategic planning takes a few weeks of dedication, in which the process can take longer according to the location of plants and involvement of teams and suppliers when necessary.

The role of strategic management is to be in control of the process continuously leveraging the strategies that must be constantly revised even though it does not always achieve the intended goals. Hoshin planning must be managed because it is an ongoing process that needs feedback. As an example and to get a glimpse on how to manage the strategies set out in the corporate, managerial and functional levels in an integrated manner and online see Figures 40 and 48.

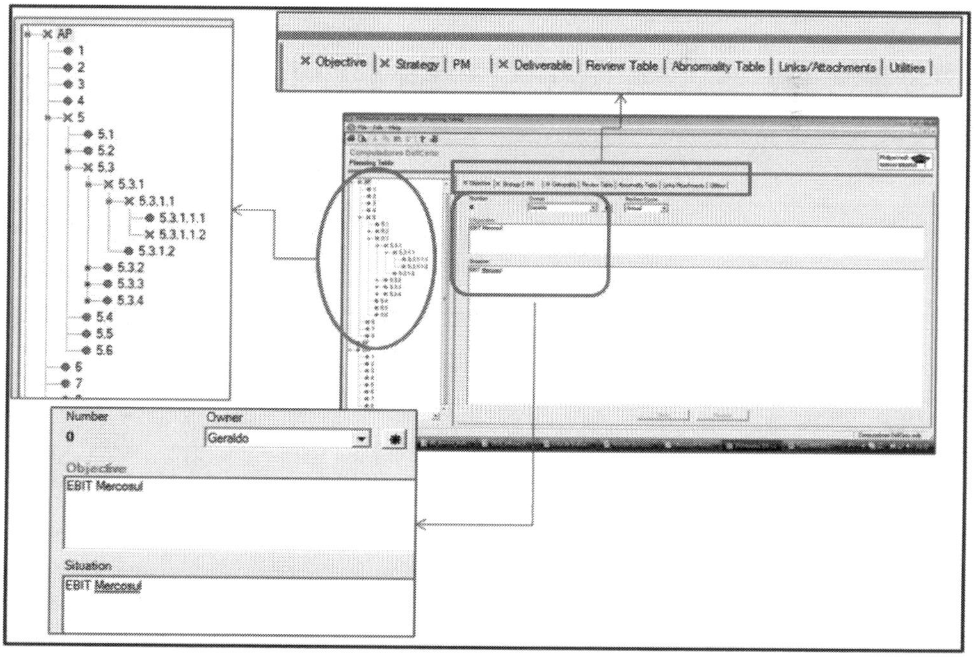

Figure 48 PC-Hoshin software screens (TQE, 2011)

Once the projects have been defined it is necessary to control and maintain them. Improvements must be controlled under a specified frequency. The plan of action and monthly graphics are important and are included in the routine management of the day-to-day. As an example you can clearly define which one of these graphics is the concept of the income statement, tips, historical and trend. A plan of action is shown for example done in Microsoft Office Excel (graph 4). This type of chart becomes interesting because it contains a visual interface for control and it already contains an action plan describing the actions, which will be responsible for them, planned and carried out impact, start time and end and the status using the nomenclature troubleshooting method - PDCA, which is easy to disseminate to all employees. According to several authors, among them Dennis (2006), the problem solving can be achieved with the use of kaizen.

Graph 4 Action plan and monthly control of OEEa

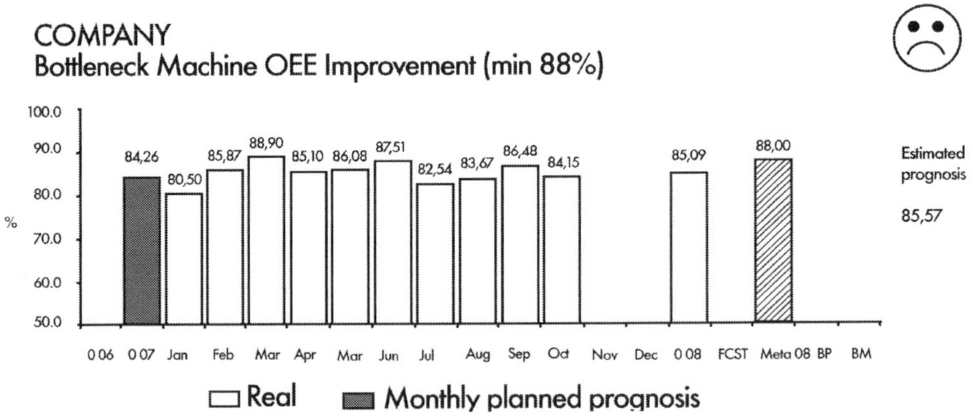

	Basic Actions / Additional Actions	Owner	Action Impact Plan	Action Impact Real	Data Start	Data End	Status
1	Preventive maintenance plan for molds	Josiel	88,00	85,09	07.01	30.12	C
2	Keep spare parts in stock	Pires	88,00	85,09	01.02	30.12	C
3	Reactivation of preheating table molds for basic and super lines	Pires	88,00	85,09	12.05	10.05	A
4	Super Line truss system reform	Pires	88,00	85,09	22.05	25.05	A
5	Preheating system for KDN Molds	Santos	88,00	85,09	01.04	05.05	A
6	Implementation of daily loss management of production due to machinery	Marcos	88,00	85,09	01.06	30.12	A
7	Implementation of daily set up controls	Marcos	88,00	85,09	01.06	30.12	A
8	Revision of pneumatic Mega Line system	Pires	88,00	85,09	07.07	11.07	A
9	Revision of OEE according to Guideline (for 2009)	Calado			30.10	30.12	P

Action Planning   Action Performance   Result measurement   Standard fixation

Visual management   Company   05/25/2011

As it was already explained you can also have a management and control software like PcHoshin software to integrate all action plans, monitor and control of actions and projects in an integrated manner, all giving results that lead to effective strategic management (Figure 48). The PcHoshin, as integrated management option, includes the example of a sketch model of the Integrated Company Diagnostic method proposed in this work (Figure 40), in which their tools, methods, more relevant strategies and success requirements and basic contents are discussed throughout the chapters of this work content.

5. ENTERPRISE DIAGNOSTIC METHOD APPLICATION

The scope of the field research is the diagnosis and identification of the strengths and weaknesses of enterprises management´s, supporting investment decisions and providing strategic information and opportunities for improvement in the processes. The diagnosis will plan actions to improve the management´s ability over the equipment and the personnel involved in the production system as well as conduct a preliminary assessment of the strategic management of operations and company Logistics.

There have been research divided into four different segments, three private companies and a public service, which are the machined parts industry segments; of molded parts industry; textile and sanitation services, although this chapter will only describe the test appliance of the EDM in one of the aforementioned companies four, because of confidentiality reasons no names are disclosed.

5.1 EDM applied in the industrial machining segment

The objective was to diagnose and identify the strengths and weaknesses of operations management and services of a machining industry, private company, located in the state of São Paulo, and will henceforth be called machining company, for reasons of business secrecy.

A medium-sized company in the mechanical metal processing segment that caters to various customers in the precision machining market; stamping mechanical serial assembly. The work subsidizes investment decisions and provides strategic information and opportunities for process improvements to business people with the same participation.

In the four companies surveyed, EDM is being used in a timely manner and will not be an instrument to compare with other companies in the same industry, making partial use of the concepts of industrial benchmarking. It is noteworthy that all confidential information obtained from the machining company that performed the EDM will be kept in strict confidence and will only be disclosed with the express documented authorization of the company.

5.1.1 Industrial scenario for mechanic metal processing

The data collected in the Statistical Yearbook of Production Development Registry is a collection of information that is intended to provide updated indicators of the industrial sector, presented according to the National Classification of Economic Activities (NACE) IBGE. This data shows the result of the crisis for mechanical metal processing sector in 2009. It follows the evolution of the industrial sector and its respective performance. The number of jobs grew, but on the other hand, other indicators such as production, capacity utilization and industrial performance fell. Productivity rose, but still significant improvement was needed (Table 3).

Given these indicators, showing is the reality in the processing industries, there had to be a reduction of the use of industrial capacity. The volume drop scenario, 2009 had a number of consequences, such as low OEE - Overall Equipment Efficiency (Overall Equipment Effectiveness) and unemployment increased and there was damage to the processors in general.

Such a scenario was gradually recovering in 2012, e.g. for machined and stamped parts manufacturers whose main customers are companies in the automotive segment, in the case of machining companies, are dependent on the governance and the automotive segment growth, main customer, i.e. It is necessary to have a significant increase in domestic consumption of motor vehicles and, if possible, to resume exports, but for this, it is ideal that the entire segment chain reaches the world-class level, globalized and sustainable.

Chart 3 General indicators of mechanical processing segment (STATISTICAL YEARBOOK, 2012)

INDICATORS	2005	2006	2007	2008	2009	2010	2011	MEAN 2005-2011
Industry Physical Production by Class and Gender Base: the same period last year = 100) - With Seasonal Adjustment Real Growth Rates (%) - Vehicles	7.1	1.4	15,0	8,1	-12.4	24.2	2.7	6.58
2. Regional Industrial Performance - Real Growth Rate (Base: Same Period Previous Year = 100) - No Seasonal Adjustment - Sao Paulo	3.7	3.2	6.2	5.2	-8,4	10.2	0.2	2.9
3. Average utilization of installed capacity Manufacturing Industry (%) - Mechanical Industry	85.4	80.4	85,0	86.6	73,5	83.4	84,9	82.74
4. Jobs numbers by Industry Genres (Formal Market) - Manufacturing Industry (x 1000)	6.008	6.253	6.710	6.905	6.908	7.517	7.681	6.854
5. Jobs numbers by Industry Genres (Formal Market) - Fab Prods.. Metal, Except Machinery and Equipment	378,865	410,868	452,830	484,123	478,148	526,202	540,540	467,368
6. Busy staff According to the activities in industry - Parts and Accessories for Motor Vehicles	288,919	310,284	349,458	345,896	337,182			326,347
7. Variation of Annual Labor Productivity for Activities in Industry (%) - Parts and Accessories for Motor Vehicles	-2.6	-6.4	0.4	2.1	-19.4			- 5.18

Note - Annual Labor Productivity Variation was estimated as the ratio of the change in value added, the previous year's prices and the variation of employed persons (Source: http://www.mdic.gov.br/arquivos/dwnl_1337260114. pdf).

The competitive condition is in fact important, at least for the four companies that were surveyed, to the operations strategy and Logistics which has an integrated strategic management (proposed model in chapter four) and a use of industrial processing capacity with superior quality and competitive costs. In today's competitive global scenario and the processing segment in constant threat, it is appropriate to implement the EDM to learn how to have a better understanding of waste and waste elimination, as well as seek new ways to add value to customers.

5.2 EDM application of planning in machining – PDCA Plan

The first phase of the EDM is considered as the PDCA problem solving method; therefore, it is oriented to start the planning of the EDM. Some actions that were structured using the action research approach as their backbone follows the order previously described in the method, we only describe the relevant steps for teaching purposes. Planning contains the first six steps: preliminary proposal of EDM research; structure of researchers; literature review; contact with the company's management; identification of participants / EDM consultants; and application of action research.

Again placing the direction of machining company as an example we clarify that 4th stage of EDM, happened through a third company, a provider of information technology, and this relationship could be scheduled since the first meeting of EDM with the direction. In this pre-scheduled meeting with the machining company, the researcher has the opportunity to introduce him or herself, as well as the opportunity to introduce the goal of his research and partnership proposal that the EDM application project.

Once you have understood the advantages, management allows an early EDM, which took place in company, and it can take 5-10 meetings. Then the management team of evaluators who are reliable individuals and usually managers of key departments are the ones who are invited to participate by high management. Evaluators receive direction via an invitation to the first day of the EDM Workshop.

The first three steps have been structured by the researcher, because the intention of the researcher can evaluate the different segments in the Campinas region to analyze, improve and EDM study hypotheses to add value, and has the already pre-defined order of the implementation of EDM.

There is a researcher and owner, because the survey methodology in several meetings happens practically within workshop in which it is necessary to know the matters pertaining to EDM and preferably experience of Lean Enterprise, Benchmarking, Hoshin Planning, Strategic Management and other knowledge.

The researcher's knowledge of EDM and of business processes usually end up generating long discussions on the issues of questionnaires during interviews with the evaluators who are experts in their own right. The importance of research and workshops moderator to ensure the comfort, creativity, cooperation and motivation to the evaluators is worth mentioning. Everyone has the opportunity to express themselves and to question every step of the EDM, strengthening the implementation of the action research method. All planning steps were performed as described in Step 6 of the EDM, the researcher gets to choose the moderation techniques, questionnaires, interviews and action research.

5.3 EDM application for machining company – PDCA Do

Continuing with the analogy PDCA/EDM problem solving method, we will discuss steps seven through eleven to run up the diagnostic workshops and thus we will discuss guides to start the implementation of the EDM.

In this 7th stage the first contact an important meeting between researchers and the participants takes place. That meeting is a consensus with the project team, therefore it cover the following aspects.

The industrial director presents the researcher to all participants and expressed formal support for the project, as there is the need to have consistent information for decision making, achieve goals and results in an integrated company, aligned, participatory and sustainable. The director and sponsor make clear the purpose of the EDM workshop.

The researcher makes a dynamic presentation to meet the evaluators, their experiences in the company and their area of expertise and responsibility. Then the researcher lectures about the competitiveness of enterprises, the difference between a conventional company and a modern enterprise, explains about the history of the Toyota Production System and Lean Enterprise until he reaches the main point, the necessity to add value to customer and reduce waste to increase the company's assets in an integrated, participatory and sustainable way.

The participants hear a brief explanation that this work is intended to support decision-making and will be critical to the Hoshin planning, and the researcher also explains that the identification of weaknesses will be judged by the participants and used to leverage and create a list of recommendations projects for improvement or maintenance.

It is clear that the EDM is participatory and collaborative. This is a self-assessment process that will result in a final report and this report will be the result of all the work to implement a strategic planning and leverage at the end with a new control model that will obtain results through strategic management. EDM aims to develop a strategic plan with synergy, with the participation of area representatives and focused on the bottom line with sustainable business vision.

In the 8th stage all participants receive the questionnaires which are returned to the researcher by the end of the EDM workshop day. The questionnaire was answered in a written manner at the same time as the researcher is explaining the issue and the concepts, participants are encouraged to attach their notes if they desire. In the case of this machining company, it is a company with culture and conventional production system; therefore, as expected, there was explanation of the concepts related issues to facilitate the learning process and knowledge of all on each question. Then all participants gave a single grade in question (1 to 5). Participants described the reason for this grade on the form as the EDM assigns grades or notes that must be followed with an explanation as to why you would qualify with that particular number or choice.

In the next step, the participants struggled many times over the assigned grade of some issues, there was a more detailed reflection on the critical issue of the company, but the intention to promote the continuous improvement of vision facilitated the group to conduct a self-analysis of where and how this can be improved. For example, discussions based on modern concepts of operations and Logistics, Lean Enterprise and Hoshin Planning were inevitable and enriching, in the opinion of the company's Logistics Machining manager.

5.3.1 Research capacity management

The Machining Company's diagnostic questionnaire also did the implementation of management ability questionnaire, struggled on the issues and actual situation of a machining company. For more information on the equipment capacity management can be consulted article "Production management through the OEE: the case of an electromechanical company", published by Calado and Calarge (2009).

The ability of management research (OEE - Overall Equipment Efficiency ((% quality x% x% performance availability) / 10,000), form attached in Technical Report shows an average OEE and estimated 68% (99.5% x 95 % x 72% = 68%). With a low OEE it is necessary to timely implementation tools such as SMED TPM or to reduce losses due to the tool change time in milling machines.

The ability management to researched questions regarding the use and time is acceptable since they apply the tools and Lean methods on the factory floor such as 5S; PDCA; ZIP CODE; standardized work; SMED - Quick tool change; autonomous maintenance; control charts; plan ideas and OEE itself, and if the evaluators have such knowledge or apply partially. It is noteworthy that the global companies already make use of TQC approaches, TPM, 6 Sigma and Lean.

The company researched, for example, has a PCPMaster software ACD company to improve the ability management, which provides online data, process improvement of information flow that has existed for some years. The production uses the OEE methodology, is also responsible for managing the OEE and during the diagnosis, it was found that the production manager uses the software when he needs operational information. It is worthy to record the monitoring and control of production capacity, the industrial management, also has an online form as if it were an audit of bottlenecks processes at any time. Maintenance uses EEO every day.

Not all managers joined the EU to explore integrated management model, and there are still areas that do not use the information available in the system, although many persons have basic information about the system currently available. They do not have a formal recognition for the improvements in OEE, although the process improvements continue to occur.

The OEE's annual target in 2010, according to industrial direction was estimated to increase in around 10% to 20%, and in 2009 there was not an explicit goal. It is believed that the principal results obtained with OEE 2009 are related to the reduction of costs and increase in production in order to improve 5% to 10%.

The evaluators describe the positive aspects of ability management and OEE online are in a position to know the actual availability of its application (machine) and helps visualize the gains. Thus, they consider waste and prioritization of actions. On the downside, they're graphing rework to meet the management's demands and there needs to be a change in software requested.

There was the opportunity to apply the techniques and methods of Total quality and Lean Production, worth assessing the prioritization of implementation of SMED and TPM and other quality improvement tools and productivity.

5.3.2 Value stream mapping application

On the last day of workshop the whole group built on the shop floor a mapping of the value stream (VSM) and monitoring the flow of materials, the result were presented, (you can refer to the book learning to see Rother and Shook (1998)). In the next step, a list of recommendations is made based on the explanation of the grades and marks awarded in the subject of performance improvement for the machining company, this information is included in the final report.

The current VSM, was held together with the evaluators on the shop floor, as the methodology recommendations, is shown in Figure 49, in which it concluded that the lead of 34 working days production team is extremely long; they also identified the manufacturing processing time has reached a little over 5 minutes, to be exactly 313 seconds. In this case it is not considered 4 days from the time the heat treatment process begins, which is conducted in another plant that belongs to the researched machining company. There was consensus among the evaluators that the production flow is conventional and includes many activities that do not add value to the customer.

Before the waste was identified in the machining company during the VSM current state (Figure 49), and taking into account that the evaluators are already familiarized to the identification of waste, due to the various meetings of diagnosis, in which they expounded on methods and Lean Enterprise tools and discussed are several requirements for improved operating performance, we identified some opportunities for improvement (waste) to be considered in a future VSM, and that identification can turn into actions or into a project: in process; standard production lot; uneven demand; standardized work; kaizen to increase availability; self-control; SMED quick tool change; 5 S; TPM; improve the information system, etc. The current VSM state allows you to design a future state including the necessary process improvements. The future VSM can be used in the strategic planning phase.

5.4 EDM Application for Machining Company –PDCA Check

The EDM uses indicators that assess the company's management in seven different areas, as it is done in industrial benchmarking (Figure 50). Assessing the current situation to achieve better performance against the marks awarded to the practices and performance (Performance). The only difference lies in the depth of the questions asked in the surveys or questionnaires and the fact that it is focused on operational area as a central function. And other areas are treated as support areas.

Figure 49 Current Machining Company Value Stream Mapping

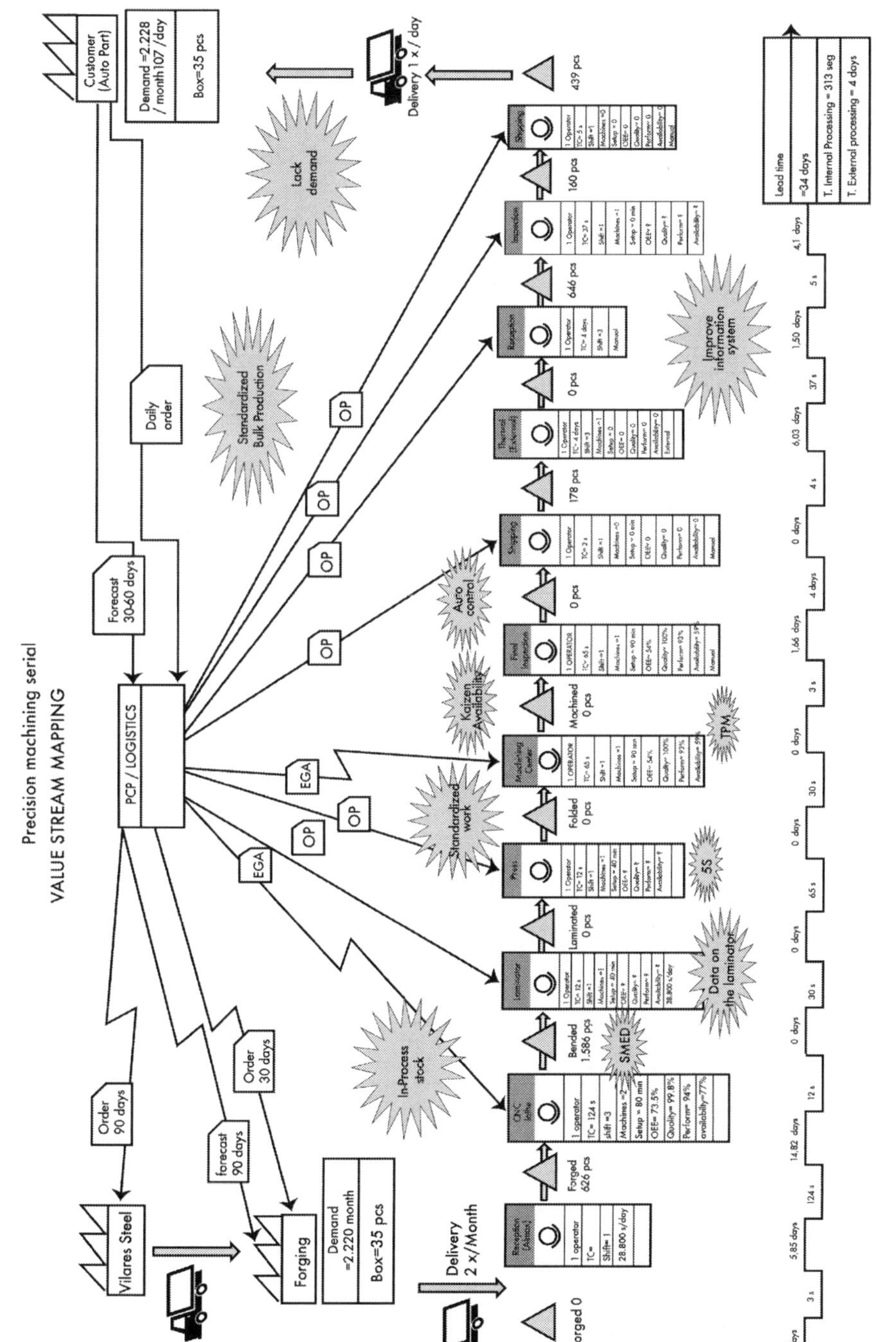

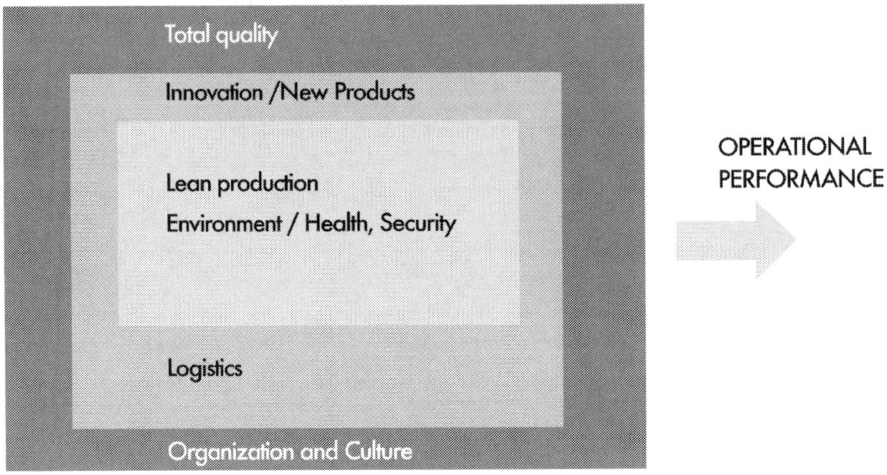

Figure 50 World Class Manufacturing Areas Model (adapted from VOSS, 1997)

Analyze practices (PR) and performance (PF) through indicators of the following areas of the company (dimensions): Total quality; Lean Production; Logistics; Organization and Culture; New product performance; Innovation management; Environment, Health and Safety. The calculations to obtain the result of EDM indicators in machining company by area (Chart 5) are in detail in the technical report.

This penultimate stage of EDM description in machining company covers the steps twelve to seventeen step, therefore, in order to first to work out the calculations we must wait after analysis of diagnosis, with Grey correlation analysis method for the selection and evaluation of the indicators. The method facilitates the analysis, selection and evaluation of the indicators and serves to facilitate understanding, planning and preparation of the final report.

To describe the application of the Grey system in the EDM, one of the areas of world class manufacturing was chosen, Lean Production area. Correlation analysis is a statistical Grey System multifactor analysis. The sample data of the factors, the degree of gray correlation is used to describe the strengths and weaknesses, the size and shape of the relationship between the factors. If the sample data reflects the same mutative situation of two factors, the degree of correlation is greater, by contrast, is lower. The combination of a set of 20 for five indicators evaluated during the management EDM are shown in Table 9. After being analyzed by gray system, it identifies that the first three strengths in the Lean Production area are the indicators AD6 = Productivity; IV1 = Average age of plant and equipment and TC1 = Time of the company's total cycle.

Chart 4 Results of EDM indicators in machining company by area

Enterprise Diagnostic Areas	%
Total Quality PR	45%
Total Quality PF	83%
Lean Production PR	53%
Lean Production PF	60%
Logistics PR	45%
Logistics PF	69%
Organization and Culture PR	52%
Organization and Culture PF	40%
New Product Performance PR	62%
New Product Performance PF	67%
Innovation Management PR	33%
Innovation Management PF	70%
Environment, Health and Security PR	65%
Environment, Health and Security PF	70%
General Performance Indicators PF	80%

5.4.1 Selection and Evaluation Method for Indicators in the Production Area

For years the characteristics of Lean Production, or Lean Production has been described by both authors, among many others regarding value the Hoshin planning, strategic planning and ways to diagnose the level of maturity of companies.

Through experience in implementing the EDM, the various indicators in the process of Lean Production (AD = Performance evaluation in the Company; IV = Investment; TC = Time of production cycles; FE = Plant and equipment; OC = Organization and Culture; Q = quality), are evaluated by the managers of the areas described in the following tables with the letter W in this case represents is evaluators (W1, W4, W5, W6 and W7), representatives of production and quality, process, human resources, Logistics and management representative of the company under study, the twenty indicators are semi-quantitative and are shown in Table 9. the W2 and W3 managers, did not participate in this stage of diagnosis, were therefore not considered.

Table 9 Rating of the Indicators - Lean Production machining company

	Indicators / Appraisers	W1	W4	W5	W6	W7
AD 4	Inventory turnover	3	3	3	3	3
AD 6	Productivity	4	4	4	4	4
AD 7	Cost of products	3	3	2	3	3
AD 8	Performance measures	3	2	2	3	3
FE 1	Layout of equipment (factory floor layout)	4	3	3	3	2
FE 10	Housekeeping/ 5 S	2	3	3	3	3
FE 2	Automation	2	2	2	2	2
FE 3	Pull Production	2	3	2	3	3
FE 4	Batch size	2	3	4	4	4
FE 5	Warehouse Truck	2	2	1	2	2
FE 8	Maintenance	2	2	2	3	2
IV 1	Average age of plant and equipment	4	4	3	4	4
IV 2	Investment Capacity	2	1	1	1	1
IV 3	Investment Return time	4	2	2	3	2
IV 4	Technological modernization investment	3	3	3	3	2
OC 6	Flexibility at work	3	3	3	3	3
TC 1	Time of the company's total cycle	3	4	3	4	4
TC 2	Material processing cycle time	2	1	3	1	2
TC 6	Equipment preparation time	3	2	4	3	3
Q 8	Relationship with Suppliers	3	2	2	2	2

Table 9 is shown as a non-dimensional shape by the linear standard method, while all contents are unified in positive index into an array x:

$$X = \begin{bmatrix} 0.750 & 0.750 & 0.750 & 0.750 & 0.750 \\ 1.000 & 1.000 & 1.000 & 1.000 & 1.000 \\ 0.750 & 0.750 & 0.500 & 0.750 & 0.750 \\ 0.750 & 0.500 & 0.500 & 0.750 & 0.750 \\ 1.000 & 0.750 & 0.750 & 0.750 & 0.500 \\ 0.500 & 0.750 & 0.750 & 0.750 & 0.750 \\ 0.500 & 0.500 & 0.500 & 0.500 & 0.500 \\ 0.500 & 0.750 & 0.500 & 0.750 & 0.750 \\ 0.500 & 0.750 & 1.000 & 1.000 & 1.000 \\ 0.500 & 0.500 & 0.250 & 0.500 & 0.500 \\ 0.500 & 0.500 & 0.500 & 0.750 & 0.500 \\ 1.000 & 1.000 & 0.750 & 1.000 & 1.000 \\ 0.500 & 0.250 & 0.250 & 0.250 & 0.250 \\ 1.000 & 0.500 & 0.500 & 0.750 & 0.500 \\ 0.750 & 0.750 & 0.750 & 0.750 & 0.500 \\ 0.750 & 0.750 & 0.750 & 0.750 & 0.750 \\ 0.750 & 1.000 & 0.750 & 1.000 & 1.000 \\ 0.500 & 0.250 & 0.750 & 0.250 & 0.500 \\ 0.750 & 0.500 & 1.000 & 0.750 & 0.750 \\ 0.750 & 0.500 & 0.500 & 0.500 & 0.500 \end{bmatrix}$$

At this point most ideal samples X0 = (1, 1, 1, 1, 1, 1, 1, 1, 1, 1, 1, 1,1, 1, 1, 1, 1, 1, 1) because of the formula (1), calculate the absolute difference of the sample X0 and Xi, and is the absolute difference matrix Δ.

$$\Delta_{ij} = \left| X_{0j} - X_{ij} \right| \quad (1)$$

Once the Δ (max.) And Δ (min.) are calculated equal ratings were assigned for all evaluators, assuming that the information the evaluators have equal importance. Not applicable for the use of the organizational hierarchy to define the degree of importance of the evaluators on indicators (indexes). There may be professional hierarchy, but no information hierarchy.

$$\Delta = \begin{bmatrix} 0.250 & 0.250 & 0.250 & 0.250 & 0.250 \\ 0.000 & 0.000 & 0.000 & 0.000 & 0.000 \\ 0.250 & 0.250 & 0.500 & 0.250 & 0.250 \\ 0.250 & 0.500 & 0.500 & 0.250 & 0.250 \\ 0.000 & 0.250 & 0.250 & 0.250 & 0.500 \\ 0.500 & 0.250 & 0.250 & 0.250 & 0.250 \\ 0.500 & 0.500 & 0.500 & 0.500 & 0.500 \\ 0.500 & 0.250 & 0.500 & 0.250 & 0.250 \\ 0.500 & 0.250 & 0.000 & 0.000 & 0.000 \\ 0.500 & 0.500 & 0.750 & 0.500 & 0.500 \\ 0.500 & 0.500 & 0.500 & 0.250 & 0.500 \\ 0.000 & 0.000 & 0.250 & 0.000 & 0.000 \\ 0.500 & 0.750 & 0.750 & 0.750 & 0.750 \\ 0.000 & 0.500 & 0.500 & 0.250 & 0.500 \\ 0.250 & 0.250 & 0.250 & 0.250 & 0.500 \\ 0.250 & 0.250 & 0.250 & 0.250 & 0.250 \\ 0.250 & 0.000 & 0.250 & 0.000 & 0.000 \\ 0.500 & 0.750 & 0.250 & 0.750 & 0.500 \\ 0.250 & 0.500 & 0.000 & 0.250 & 0.250 \\ 0.250 & 0.500 & 0.500 & 0.500 & 0.500 \end{bmatrix}$$

The conditions for evaluators have been leveled during the evaluation workshops to ensure the criteria of EDM. The weight in this case is the degree of importance of the information that is provided by the evaluators previously selected by the direction in which sponsors the application of EDM.

Finally it assigns weights and is defined between zero and one, as a variable which belongs to the real numbers in the interval (0, 1) so that the sum of the weights equals one, and the value of one equals 100%. This condition is applied to all the case studies, and is part of the EDM.

In this case the Δ (max.) = 0.750, Δ (min.) = 0. In the evaluation strategy is adopted equal weights to the evaluators a flush time is known the weight of the various indices (wi): 0.2, 0.2, 0.2, 0.2, 0.2. Compute the correlation coefficient to the following processing of the data in the absolute difference ε matrix (2).

$$\varepsilon_{ij} = \frac{\Delta(min) + \rho\Delta(max)}{\Delta_{ij} + \rho\Delta(max)} \qquad (2)$$

The coefficient to distinguish whether the value between 0.1 and 0.5, ρ takes 0.3, you receive the correlation coefficient matrix of ε:

$$\varepsilon = \begin{bmatrix} 0.474 & 0.474 & 0.474 & 0.474 & 0.474 \\ 1.000 & 1.000 & 1.000 & 1.000 & 1.000 \\ 0.474 & 0.474 & 0.310 & 0.474 & 0.474 \\ 0.474 & 0.310 & 0.310 & 0.474 & 0.474 \\ 1.000 & 0.474 & 0.474 & 0.474 & 0.310 \\ 0.310 & 0.474 & 0.474 & 0.474 & 0.474 \\ 0.310 & 0.310 & 0.310 & 0.310 & 0.310 \\ 0.310 & 0.474 & 0.310 & 0.474 & 0.474 \\ 0.310 & 0.474 & 1.000 & 1.000 & 1.000 \\ 0.310 & 0.310 & 0{,}231 & 0.310 & 0.310 \\ 0.310 & 0.310 & 0.310 & 0.474 & 0.310 \\ 1.000 & 1.000 & 0.474 & 1.000 & 1.000 \\ 0.310 & 0.231 & 0.231 & 0.231 & 0.231 \\ 1.000 & 0.310 & 0.310 & 0.474 & 0.310 \\ 0.474 & 0.474 & 0.474 & 0.474 & 0.310 \\ 0.474 & 0.474 & 0.474 & 0.474 & 0.474 \\ 0.474 & 1.000 & 0.474 & 1.000 & 1.000 \\ 0.310 & 0.231 & 0.474 & 0.231 & 0.310 \\ 0.474 & 0.310 & 1.000 & 0.474 & 0.474 \\ 0.474 & 0.310 & 0.310 & 0.310 & 0.310 \end{bmatrix}$$

To calculate the degree of correlation (r) using the equation (3):

$$r_i = \sum_{j=1}^{16} w_j \varepsilon_{ij} \qquad (3)$$

Then calculating the r: r1 = 0.4737; r2 = 1.0000; r3 = 0.4410; r4 = 0.4083; R5 = 0.5463; r6 = 0.4410; r7 = 0.3103; r8 = 0.4083; R9 = 0.7568; r10 = 0.2944; r11 = 0.3430; r12 = 0.8947; r13 = 0.2467; r14 = 0.4809; r15 = 0.4410; r16 = 0.4737; r17 = 0.7895; r18 = 0.3112; r19 = 0.5463; r20 = 0.3430, according to table 11. The set of following the evaluation of the indicators is: AD6> IV1> TC1> FE4> (FE1 = 6MWT)> IV3> (AD4 = OC6)> (AD7 = FE10 = IV4)> (AD8 = FE3)> (FE8 = Q8)> TC2> FE2> FE5> IV2.

After the calculations and matrices on EDM assessments, in which Grey detailed implementation of the system in the field of lean production it was possible to analyze the strengths and weaknesses, as well as the analysis and understanding of the evaluation and due justifications or explanations of the grading of area indicators.

5.4.2 Strengths and weaknesses of the evaluation of the production area

According to Graph 5, which shows the average results of the assessment in the judgment of the evaluators therefore quantitatively set itself notes 1-5 for indicators of practices and performance of Lean Production. The evaluation chart and all calculations of Grey system were generated through a spreadsheet Microsoft Office Excel to avoid errors and maintain a standard in the application of EDM.

Once evaluated and having set the notes apply to Grey correlation analysis we were able to describe the correlation in table 10 (r) and shows that the AD6 indicators; IV1; TC1 are those with the best three results in the size of production.

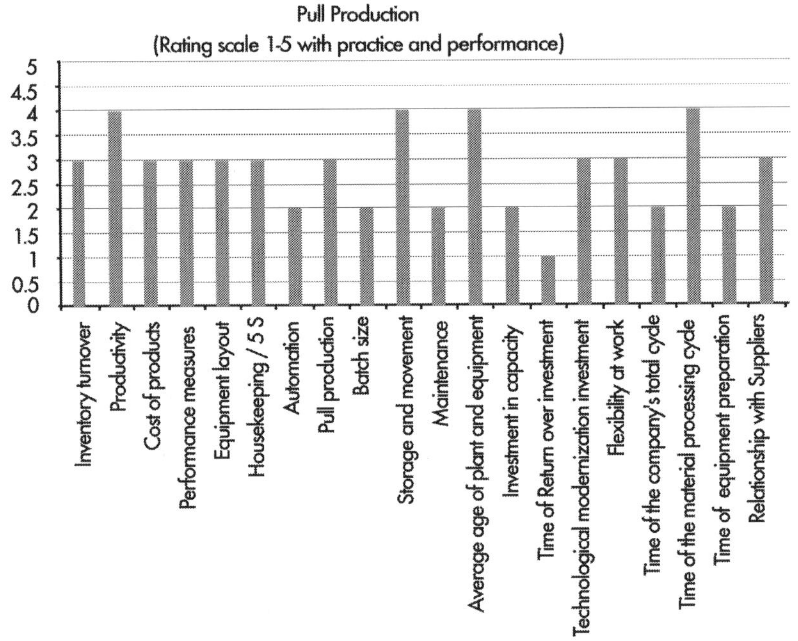

Graph 5 Evaluation of Lean Production Machining company

Depending on the degree of correlation with the evaluation indicators for the twenty Lean Production of EDM, can know that "Productivity; Average age of plant and equipment; Time of the company's total cycle" are the strengths for the managers responsible of five key areas of business.

On the other hand, the indicator "IV2 - Investment in capacity" followed by the indicators "FE5 - Warehouse Truck; FE2 - Automation; TC2 - Time material processing cycle; Q8 - Supplier Relationships "are weaknesses and deserve priority on the necessity to draw up an improvement plan. Based on records and discussions during the process of EDM, one can describe some considerations on the weaknesses: Invest in the ability (indicator IV2), but requires greater use of existing software: PCPMaster and LOGIX. The storage and handling (indicator FE5) are not efficient, more than 20% of the plant area is occupied by in-process inventories, raw material and finished products.

Table 10 Correlation of indicators - Lean Production machining company

	W1	W4	W5	W6	W7	Ri
AD 4	0.095	0.095	0.095	0.095	0.095	0.4737
AD 6	0.200	0.200	0.200	0.200	0.200	1.0000
AD 7	0.095	0.095	0.062	0.095	0.095	0.4410
AD 8	0.095	0.062	0.062	0.095	0.095	0.4083
FE 1	0.200	0.095	0.095	0.095	0.062	0.5463
FE 10	0.062	0.095	0.095	0.095	0.095	0.4410
FE 2	0.062	0.062	0.062	0.062	0.062	0.3103
FE 3	0.062	0.095	0.062	0.095	0.095	0.4083
FE 4	0.062	0.095	0.200	0.200	0.200	0.7568
FE 5	0.062	0.062	0.046	0.062	0.062	0.2944
FE 8	0.062	0.062	0.062	0.095	0.062	0.3430
IV 1	0.200	0.200	0.095	0.200	0.200	0.8947
IV 2	0.062	0.046	0.046	0.046	0.046	0.2467
IV 3	0.200	0.062	0.062	0.095	0.062	0.4809
IV 4	0.095	0.095	0.095	0.095	0.062	0.4410
OC 6	0.095	0.095	0.095	0.095	0.095	0.4737
TC 1	0.095	0.200	0.095	0.200	0.200	0.7895
TC 2	0.062	0.046	0.095	0.046	0.062	0.3112
TC 6	0.095	0.062	0.200	0.095	0.095	0.5463
Q 8	0.095	0.062	0.062	0.062	0.062	0.3430

Automation (indicator FE2) has a low level and the production is pushed, not pulled in order to produce on schedule. The time of material processing cycle (indicator TC2) is large due to external processing problems, problems with the lead time and shipping suppliers.

Regarding the relationship with Suppliers systems (indicator Q8) there must be an identification of gaps depending on corrective actions that are not used in full. There are methods and tools to improve these indicators, but are not used preventively in this conventional system of production and services, go "put out fires" and has priority to the maintenance of the technical and administrative processes.

As a second example, after the calculations and matrices on the EDM assessments; in which there was a detailed application of the Grey system using spreadsheet Microsoft Office Excel; in the area of Total quality the analysis of the strengths and weaknesses was possible as well as the analysis and understanding of the assessment and due justifications of area indicators.

5.4.3 Strengths and weaknesses of the quality assessment area a

The Lean Production is shown in Graph 6 with an average result of the evaluation according to the judgment of the evaluators. They also attributed quantitatively notes from 1-5 for the indicators and performance practices of Total quality.

According to Grey results of correlation, the analysis (Table 11) shows that if Q5 indicators; Q9; Q10 are those with the three best performing dimension Total quality. According to the degree of correlation with the evaluation of the sixteen indicators for Total quality of EDM, you can know that "warranty costs; Quality initial production of a new product, scrap costs, rework and recycling" are the strengths for managers and evaluators and staff responsible for the relevant areas of the company. On the other hand, the OC9 indicator - Customer orientation "followed by the indicators of" OC7 - Systematic use of diagnosis, OC10 - Tools for troubleshooting and Q6 - Defects (internal) "are weaknesses and deserve priority on the need to development of an improvement plan."

Graph 6 Evaluation of the Total quality of machining company

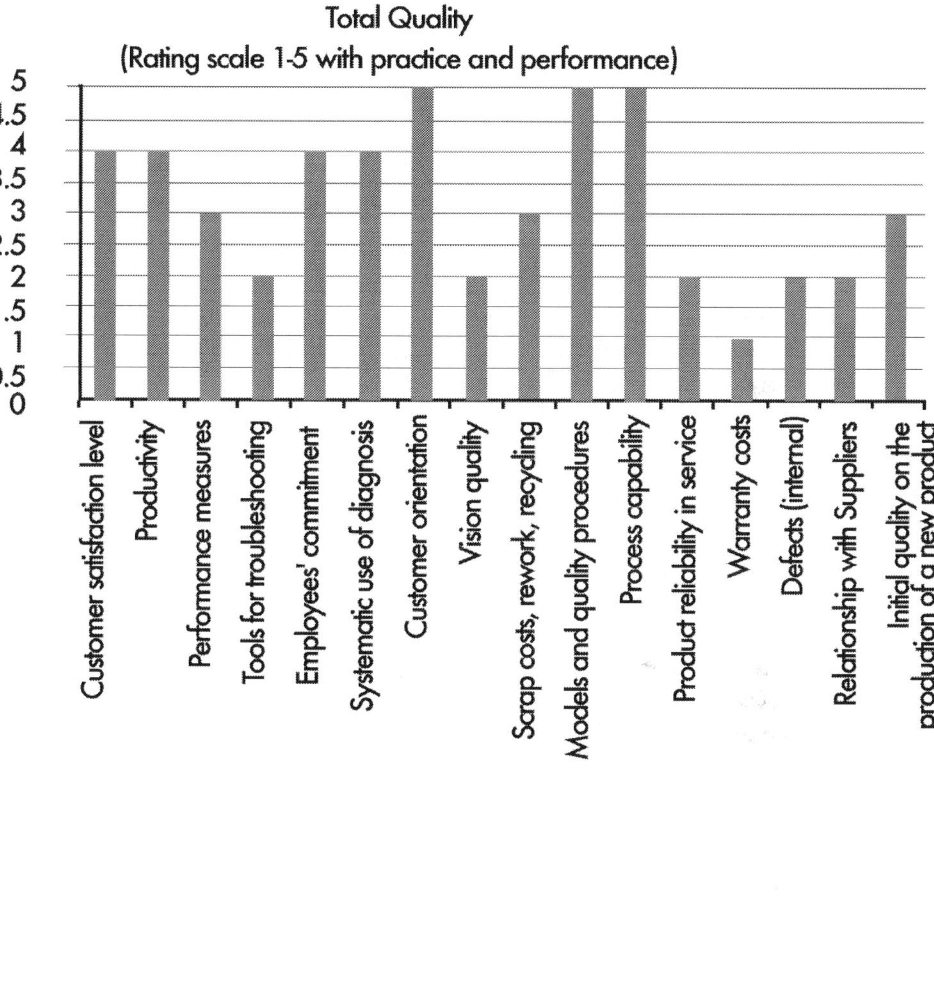

Table 11 Evaluation of indicators - Total quality of machining company

	W1	W4	W5	W6	W7	ri
AD 1	0.109	0.109	0.109	0.109	0.109	0.5455
AD 6	0.109	0.109	0.109	0.109	0.109	0.5455
AD 8	0.075	0.057	0.057	0.075	0.075	0.3393
OC 10	0.057	0.057	0.046	0.057	0.075	0.2926
Q 3	0.109	0.109	0.109	0.109	0.109	0.5455
Q 4	0.109	0.109	0.109	0.109	0.109	0.5455
Q 5	0.200	0.200	0.200	0.200	0.200	1.0000
Q 6	0.057	0.057	0.057	0.057	0.075	0.3036
OC 5	0.057	0.075	0.075	0.075	0.075	0.3571
Q 9	0.200	0.200	0.200	0.200	0.200	1.0000
Q 10	0.200	0.200	0.200	0.200	0.200	1.0000
OC 7	0.046	0.046	0.057	0.057	0.057	0.2637
OC 9	0.057	0.046	0.057	0.046	0.046	0.2527
Q 1	0.109	0.075	0.057	0.057	0.057	0.3555
Q 2	0.075	0.075	0.057	0.057	0.057	0.3214
Q 8	0.075	0.109	0.075	0.075	0.075	0.4091

Based on records and discussions during the process of EDM, one can describe some considerations on the weaknesses:

The customer orientation (indicator OC9), lacks a program to encourage creativity, greater participation and feedback, but now the formation of improvement team begins; systematic use of diagnosis (indicator OC7) is not a practice. The mission and vision of the company are only aware of the management; The tools for troubleshooting (OC indicator 10) are little used and disclosed in the company; internal defects (Q 6) are around 2160 PPM and happen internally (+ / - 2%), but do not reach the customer; as to quality procedures (Q 2) are already recognized by the team that knows the need to better apply the ISO 9001 in the factory and also extend the production benefits of search TS 9000.

The group believes that the quality of providers is not yet satisfactory; although many suppliers have good relationships. It is understood that a possible path is to improve their partnerships with them. There are methods and tools of quality management and to programs such as Total Quality Management (TQM, Total Quality Management), Lean, Six Sigma, Total Control quality (TQC, Total Quality Control) that can be used to improve these indicators, but are not preventively used in this conventional system of quality products and services, and the priority is the maintenance of technical and administrative processes, although there are several actions and projects for specific improvements.

5.4.4 Evaluation of other diagnosis dimensions in machining company

Now that the results of the diagnostic indicators are known in the company by area, the areas of the company (dimensions) of Lean Production and Total quality and stained, the Grey correlation method for selecting the evaluation of production scale was presented. The graphics are presented (Graphics 7, 8, 9, 10 and 11) in which it shows the evaluation by the majority, are evaluated quantitatively from 1 to 5 and practical indicators of performance Logistics dimensions; Organization and Culture; New product performance; Innovation management and Environment, Health and Safety.

Graph 7 Evaluation of Organization and Culture of Machining company

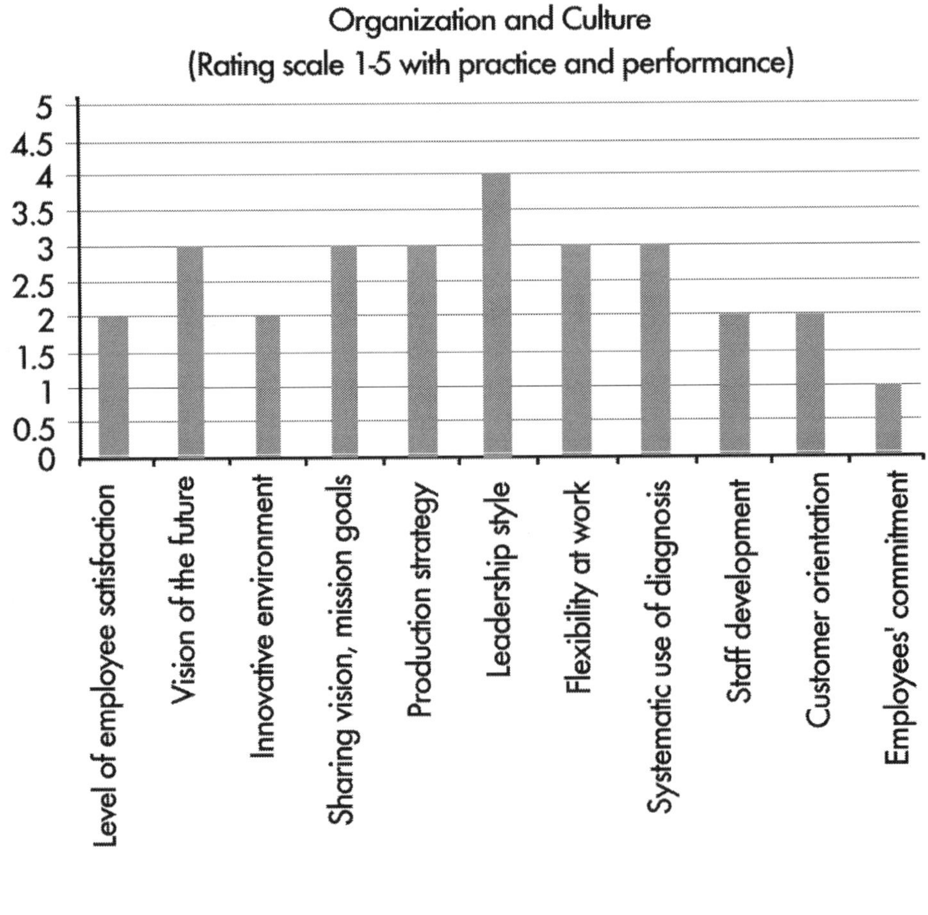

Graph 8 Evaluation of Environment, Health and Safety Machining company'

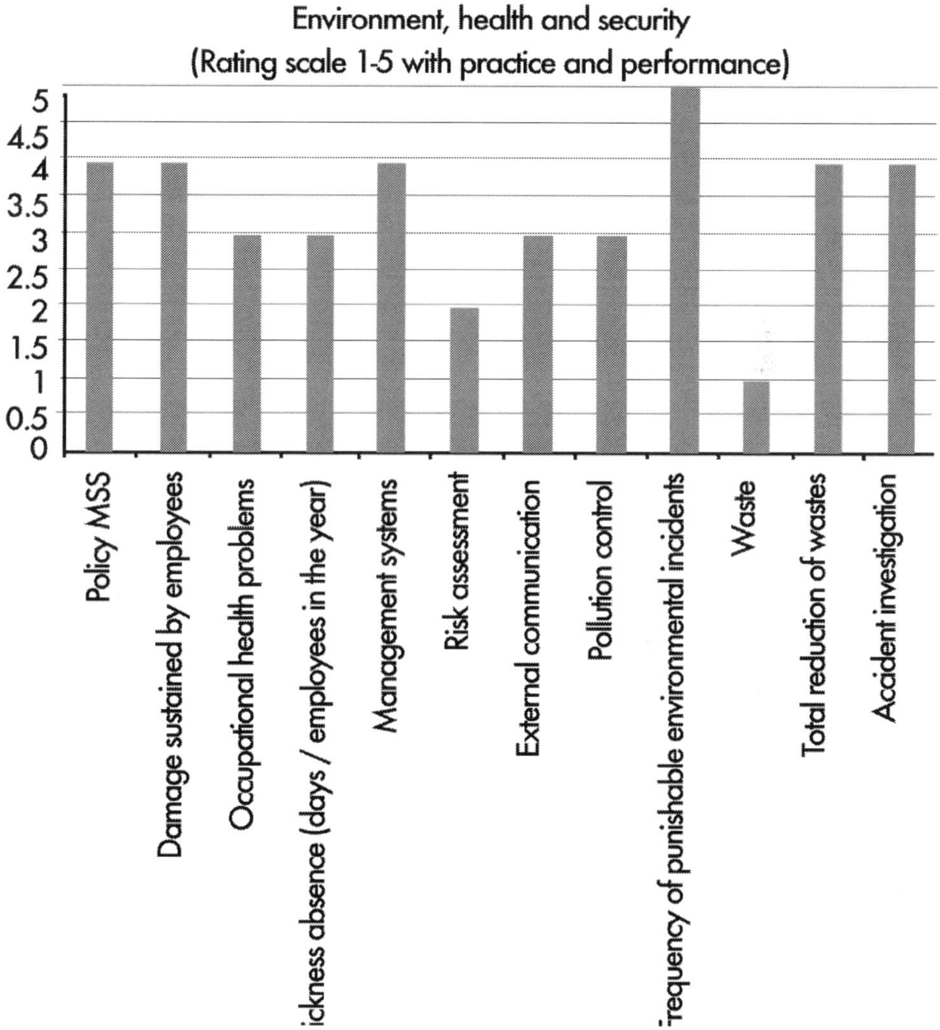

Graph 9 Logistics Evaluation of Machining company

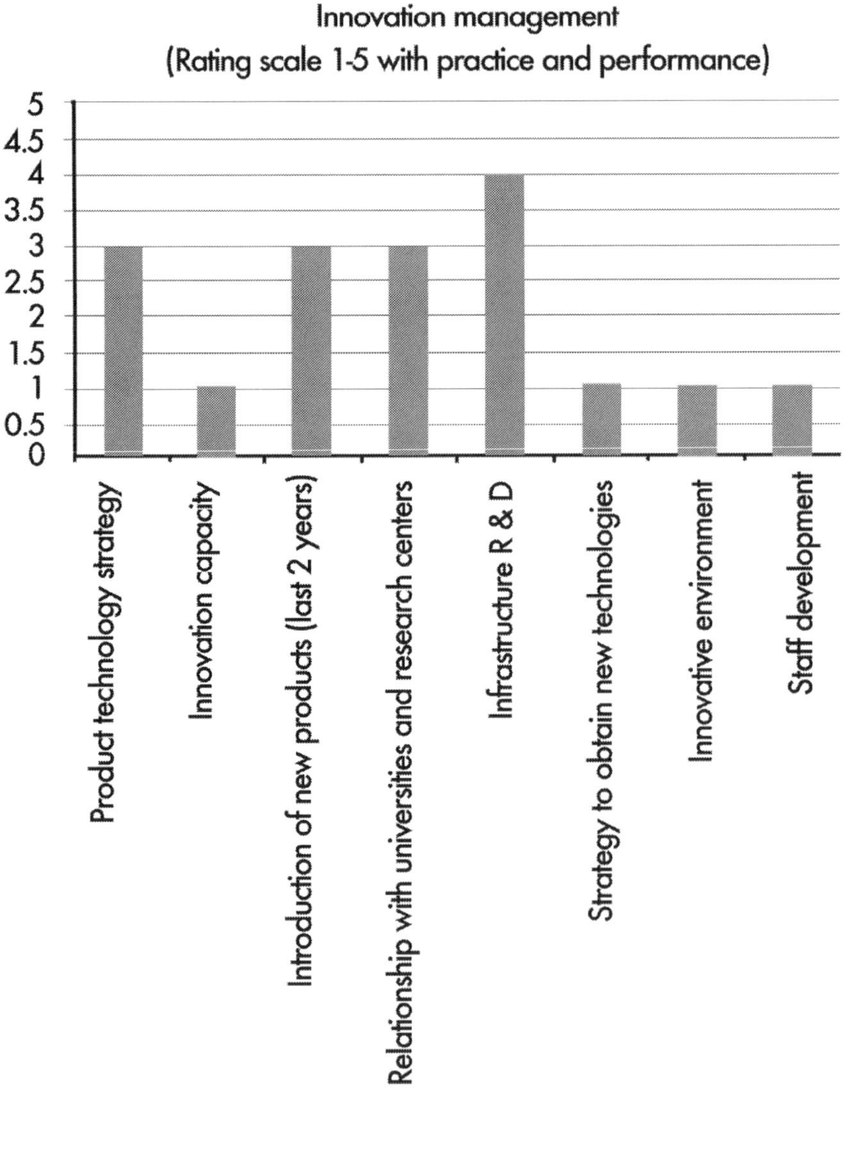

Graph 10 Innovation Management Assessment of Machining company

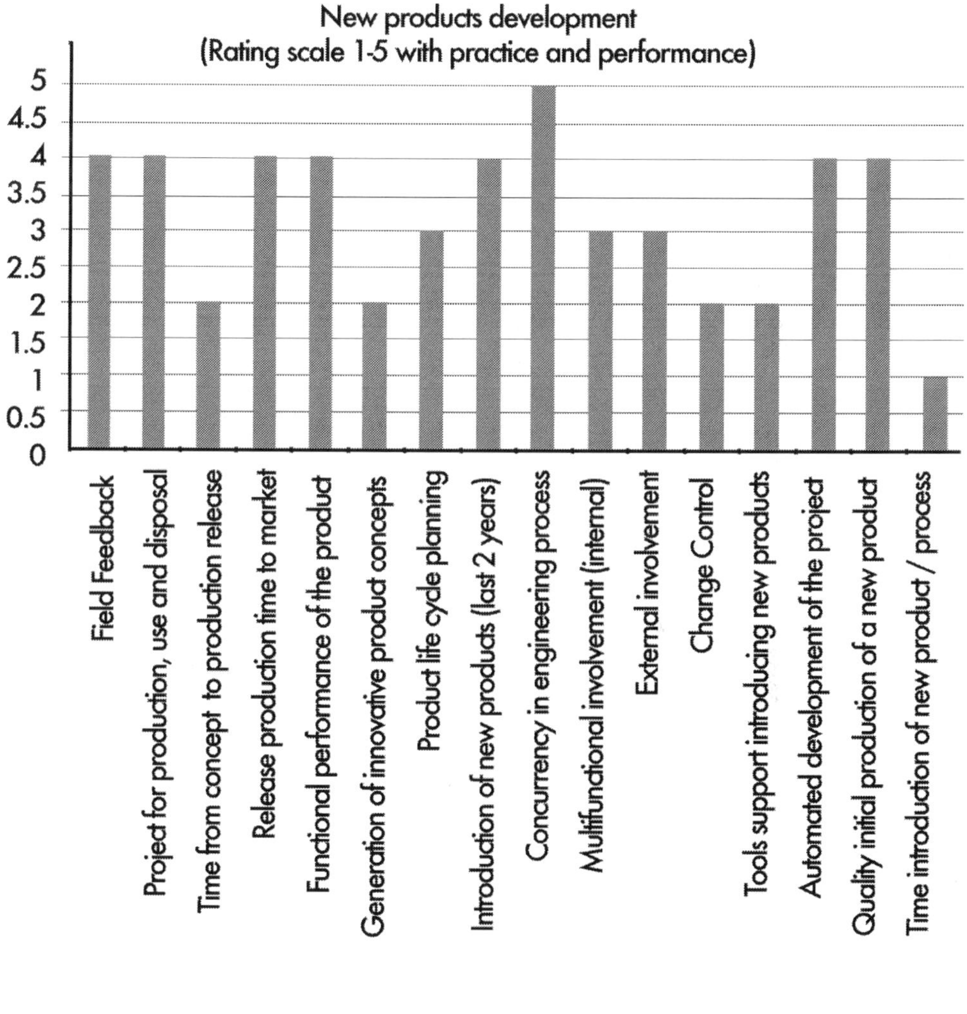

Graph 11 New product performance Machining company

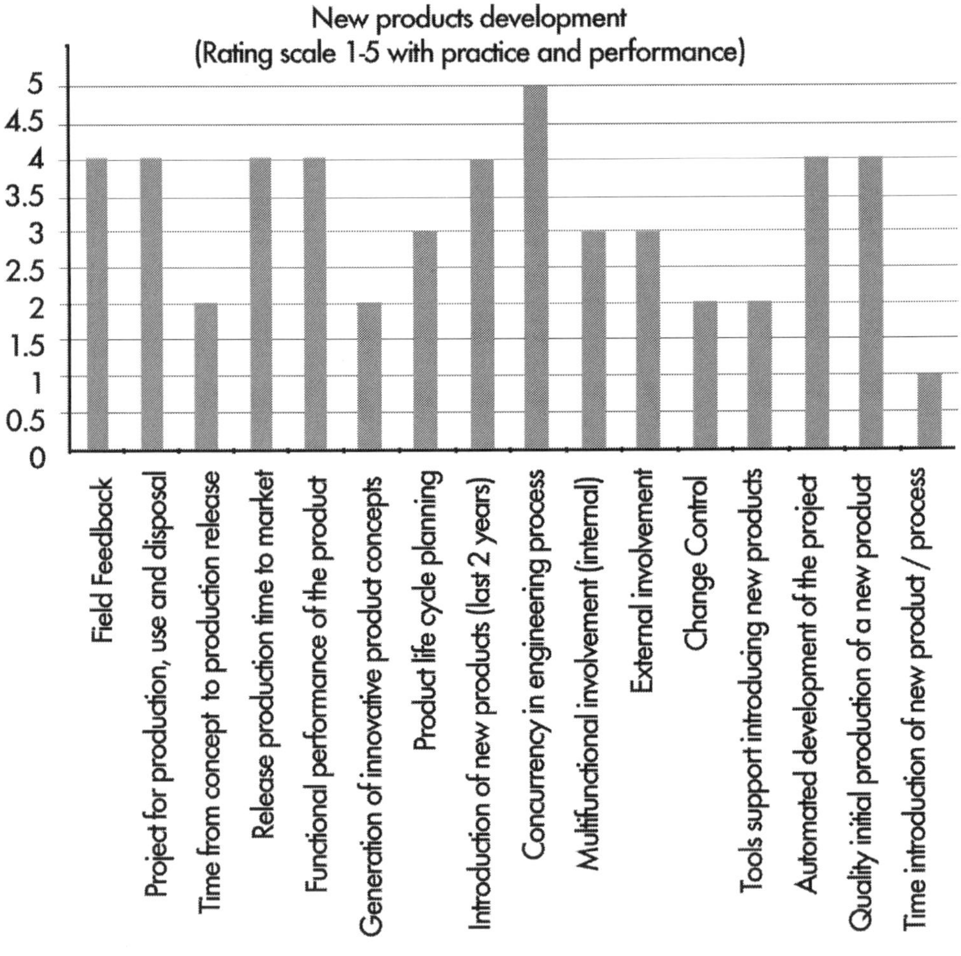

5.4.5 EDM indicators for machining company

Through the trend graph display there is the company's position in relation to the practice indices (PR) and performance (PF). The chart-radar positions the company for each of the areas evaluated in the diagnosis, in terms of practices and performances. The graph 12 has 14 axes, which represent the practice and performance indexes in each of the seven areas. Each axis has a range of 0% to 100%, and the company's position is defined by a point on this scale, totaling 14 points arranged in a circle, which are connected by lines to form a closed polygon of 14 sides and 14 vertices.

Looking at the chart-radar, one can compare the performance of Machining company under study in each one of the areas evaluated in the diagnosis. The areas where the company scored more approaches, reaches or exceeds the score of leaders are considered the strengths of the company. The points at which the distance is greater are the weaknesses of the company. Since resources are limited, it is recommended to focus efforts in taking advantage of the strengths, without necessarily prioritize investments in areas where the company has reached a standard of excellence.

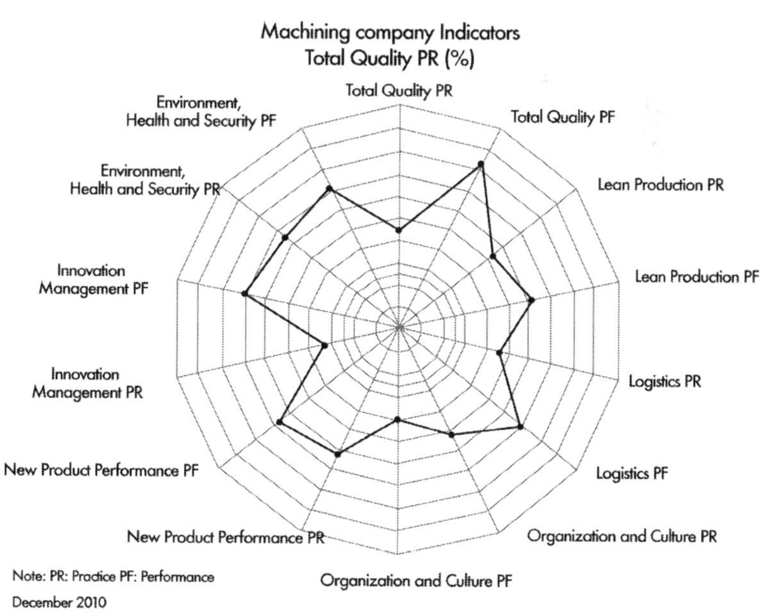

Graph 12 Indicators Machining Company EDM

5.4.6 Machining company rating

This rating allows the classification of the company according to the boxing match analogy, given in benchmarking to establish the level of maturity of the company. It appears that the real opportunity for improvement is within weaknesses, i.e. where the company is further from the excellence. The area formed between the graph lines is called room for improvement. Using this information, the company can return their focus to the areas in which improvement action will be effective, that is, produce real advances in the competitive position of the same. The machining company is in the Vulnerable category (Figure 51), as obtained in the general indicator of Practice 52% and overall indicator Performance 69%.

The methodology in its entirety is considered. The researched company will challenge, achieve a new classification in challenging category, so they will have to get an improvement of 8% in the general evaluated total in order to move from 52% to 60% in the general indicators Practice. Anyway according to the methodology the company, is far from World Class level that is above 80%. The researched company is also not performing well compared to other segments, (Figure 52), but is in a prime position to its competitors and has significant and loyal customers. We recommended a cost reduction strategy coupled with the growth strategy, identification of business principles, the implementation of Hoshin planning and strategic management. .

Getting the notes on the industrial benchmarking database IEL, and shown in Figure 52, we compare between the company and the average worldwide industry leaders. Although it can be used to compare companies in the same segment, in the case of EDM this is not the main focus of comparing companies, but the identification of the classification (ranking), the selection and evaluation of indicators, i.e., the identification of strengths and weaknesses.

Figure 51 Machining Company Rating

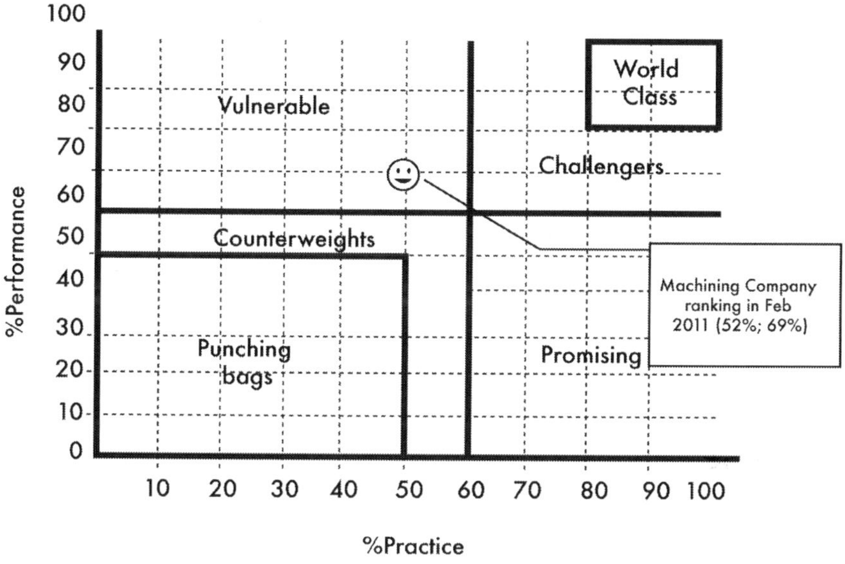

Figure 52 Rating of 3 different segments from machining company (adapted from IEL / CTI, 2010)

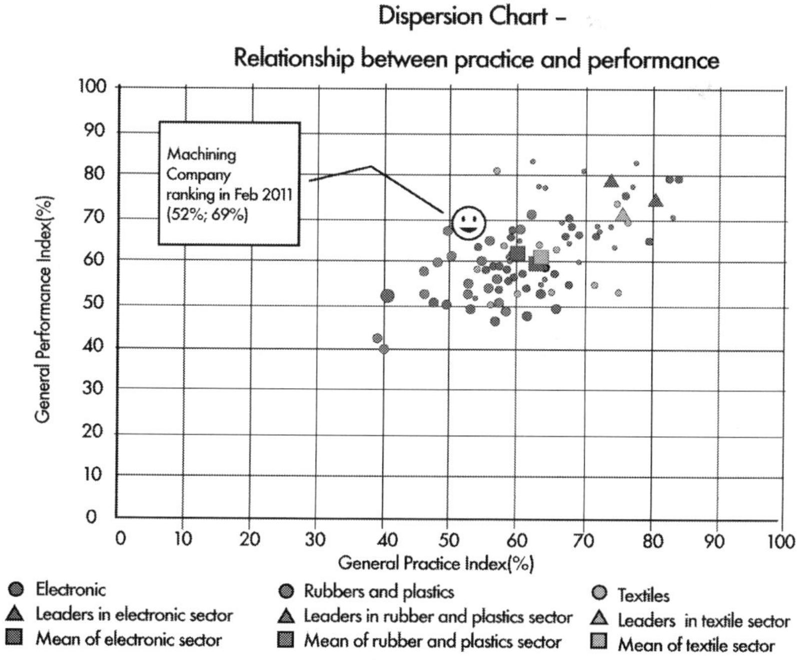

5.4.7 Main strengths of machining company

The issues in the thirteenth stage of the EDM that were evaluated with high performance were selected and given an elaborate diagnosis after the analysis of the data on maintenance of the organization, according to the Grey correlation analysis method. We can find the Grey correlation values using the Minitab statistical software Boxplot to provide quantitative information on the chart (13). The other histogram graphs (14) and confidence interval (15), aid to better understand and interpret the data.

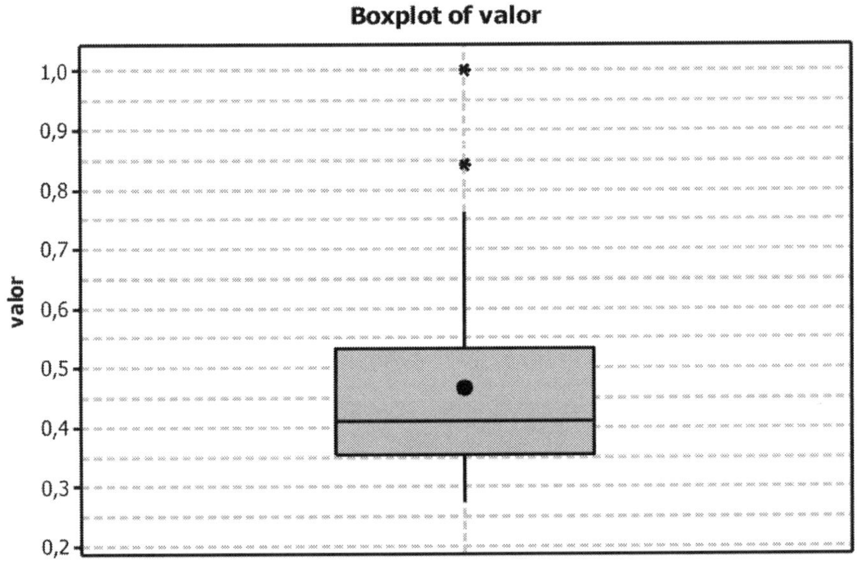

Graph 13 the correlation Box plot (ri) of Machining company indicators

The EDM propose the cut line of the first quartile (Q1) and third quartile. The machining company results show it now on the first quartile (Q1) equal to 0.352, separates the bottom 25% of the ordered values of 75% greater, which establishes a cutting line for the data set, thus identifying the worst indicators according to the first quartile.

The second quartile (Q2) equals 0.409 being the same as the median. In the third quartile (Q3) with a value of 0.531, separates from the lower 75% of the ordered values of the 25% higher thus identifies the best indicators of this dataset and results on the definition of the higher cut line.

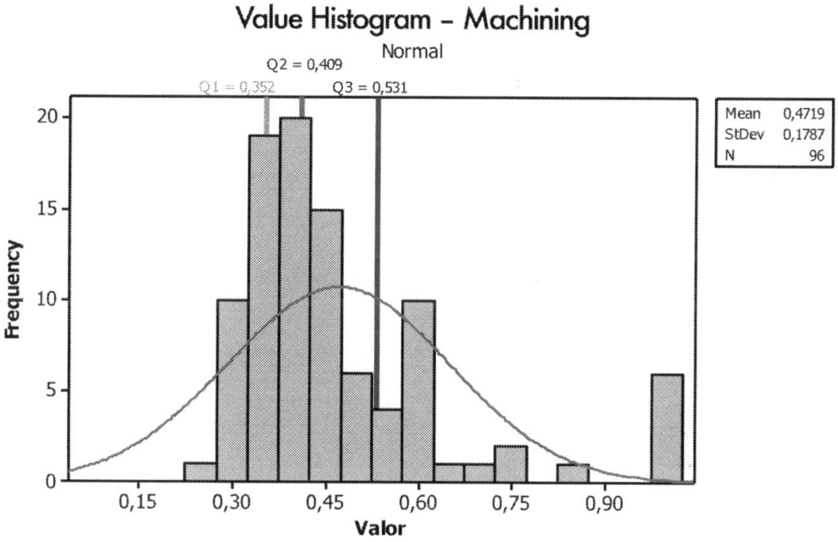

Graph 14 Histogram correlation (ri) of the Machining Company indicators

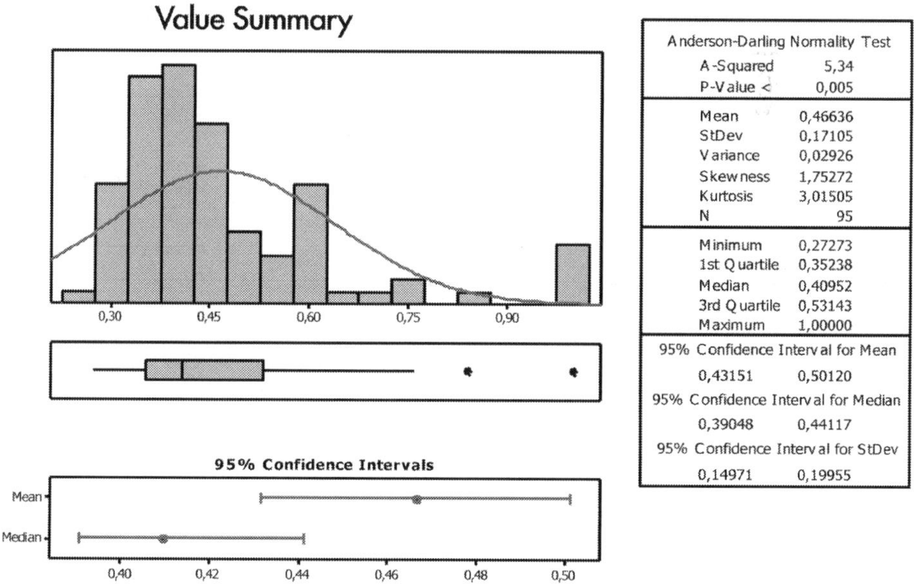

Graph 15 Summary correlation (ri) of Machining company indicators

Table 12 contains the first strengths practices and performance, according to the ordering of correlation (ri) of Grey system and classification of boxplot concept, taking into account all the evaluated indicators, are shown in Table partial reviews of each indicator . It is understood that with this list you can prepare maintenance plans (SDCA) and/or improvement projects, so as to structure the maintenance of the company's strategic plan and keeping good business results.

Table 12 The main strengths of the company Machining (3rd quartile)

Id	ri	n	Index Description
Q 9	1,00000	r15	Initial production quality of a new product
Q 7	1,00000	r32	Delivery percentage within term
MS 6	1,00000	r46	Frequency of reported environmental incidents
Q 10	1,00000	r89	Scrap costs, rework, recycling
Q 5	1,00000	r93	Warranty costs
Q 9	1,00000	r96	Initial quality of a new product
TC 3	0,84000	r35	Customer Order Processing Time
IN 5	0,76000	r8	Introduction of new products (last 2 years)
DP 1	0,68615	r1	Field Feedback
AD 5	0,64571	r26	Cash flow
DP 2	0,60615	r2	Design for production, use and disposal
DP 5	0,60615	r5	Functional performance of the product
FE 9	0,60000	r31	Change priority of production orders
MS 10	0,60000	r39	Damage sustained by employees
MS 12	0,60000	r41	Sick leave (days/employees in the year)
AD 6	0,60000	r62	Productivity
AD 1	0,60000	r81	Customer satisfaction level
AD 6	0,60000	r82	Productivity
Q 3	0,60000	r91	Process capability
Q 4	0,60000	r92	Product reliability in service
IV 1	0,56571	r72	Average age of plant and equipment
AD 2	0,56571	r25	Market share
TC 1	0,53143	r77	Time of the company's total cycle

Using the boxplot concept criteria, early 3rd quartile, the measure separatrix score statistic means that 75% of the items that meet the greater than or equal values and sets a cutoff score point is the correlation value (ri) equal to 0.53143, according to calculations that are made with the aid of Minitab 14 software, thus sets up the best indicators for machining company (table 12).

5.4.8 Main weaknesses of machining company

The questions are selected in the fourteenth stage of the EDM, evaluating underperformance according to Grey correlation analysis method. Elaborate diagnosis after analyzing the data. These are the opportunities for improvement to inform and plan the continuous improvement within the organization.

Table 13, the weaknesses of the practices and performance according to the correlation (ri), the Grey system and classification of boxplot concept, taking into account all the indicators evaluated. For a better understanding of the indicators are described in Table partial are reviews of each indicator. It is understood that, with this list, you can prepare action plans (PDCA) and or improvement projects, and improve the structure of the company's strategic plan implementations to be more competitive and achieve better operational performance, aiming at a world-class company.

Using the boxplot concept criteria, early 1st quartile, the measure separatrix score statistic that means 25% of the items that meet the values less and sets a cutoff score point is the correlation value (ri) equal to 0.35238, according to calculations that are made with the aid of Minitab 14 software, thus sets up the worst indicators for machining company (table 13).

5.5 Standardization of EDM in machining company – PDCA action

In this first stage of standardization and final research phase, step eighteen twenty-four present the results and lever up the internal environment for the annual strategic planning based on EDM, details below:

Table 13 The main weaknesses of the Machining company (1st quartile)

Id	ri	n°	Index Description
FE 7	0,35238	r30	Information systems
OC 8	0,35238	r58	Staff development
FE 8	0,35238	r71	Maintenance
Q 8	0,35238	r80	Relationship with Suppliers
Q 6	0,35238	r94	Defects (internal)
OC 11	0,34719	r23	Innovative environment
OC 7	0,34026	r57	Systematic use of diagnosis
OC 10	0,34026	r84	Tools for troubleshooting
OE 3	0,34026	r11	External involvement
OE 4	0,34026	r12	Change Control
OC 8	0,33506	r24	Staff development
FE 3	0,33333	r28	Pull production
TC 4	0,33333	r36	Delivery performance of suppliers
AD 3	0,33333	r50	Level of employee satisfaction
FE 2	0,33333	r67	Automation
OC 11	0,32814	r52	Innovative environment
TC 2	0,32814	r78	Material processing cycle time
FE 5	0,32121	r70	Warehouse Truck
OC 7	0,30909	r86	Systematic use of diagnosis
MS 5	0,30070	r45	Pollution control
TC 5	0,29697	r16	Introduction time for new products/processes
TC 5	0,29697	r37	Introduction time for new products/processes
OC 9	0,29697	r59	Customer orientation
OC 9	0,29697	r87	Customer orientation
IN 8	0,28485	r22	Strategy to obtain new technologies
MS 8	0,28485	r48	Total reduction of waste
IV 2	0,28485	r73	Capacity investment
IN 6	0,27273	r20	Relationship with universities and research centers

Before the final presentation of the results and before being taken to the official director there is, in this case, the need to redirect the job and already consider the EDM as a product that leverages best decision-making and the aggregate value for the company.

The researcher must be careful to systemically plan the presentation according to the search results, therefore, the problem that prompted the investigation is rescued and the results obtained are confronted with the (s) theory (s) that has (Ram) support research.

In the last stage of the EDM there must be formal meeting with management and guests, in this case, all evaluators who attended the closing meeting.. Management receives a technical report (final report) prepared by the researcher and owner. The names are reviewed by the researcher and other documents are provided electronically for possible systemic accuracy. A preview of strategic planning is discussed with management.

5.5.1 Recommendations for the machining company

Recommendations rclate to the ability of management research and map the value stream, fully applicable to the machining company, where the following analysis or improvements can be reviewed and recommended:

-An average company needs continuous improvement programs to increase competitiveness and make more requests;

-There is no formal long-term capacity strategy, but there needs to be better planning and ability management, as demand fluctuations may compromise the objectives as costs, working capital, speed, quality, reliability and flexibility.

-Information management in the process is manual and should be electronic on the shop floor.

-The level of standardization of methods and procedures in the factory lacks an industrial engineering strong work and quality assurance.

-There is possibility of improving performance if there are systemic actions to improve the OEE that focus on the bottlenecks and availability indicator.

Capacity management or ability management can be improved and will bring a result for the organization. Recommend to improvements in industrial controls for decision making.

Opportunity: Expanding OEE online to reduce waste in the process by managing the production capacity of main engines. Improve the system of production and services, adopting the techniques and methods Services and Lean Manufacturing (Lean Enterprise) to explore the ability and creativity of human resources and leadership.

Recommendations to the Enterprise Diagnostic survey are also fully applicable in the machining company, where you can review and recommend the following analyzes or improvements:

Implement cost reduction strategy coupled with the growth strategy through workshops with management team.

Implement the strategic deployment, set goals and objectives for all areas in an integrated and transparent way, if possible, with software support to be set.

Implement the management of routine day-to-day, starting with the work pattern of critical activities and processing bottlenecks.

Implement project 5S and visual management in processes.

Develop the mapping of the current and future value stream to guide improvements.

Implement the autonomous maintenance, leverage the TPM, to reduce dwell times and involve operational staff.

Develop training plan for managers and employees to implement the Lean Enterprise.

Leverage cost reduction and partnerships in the management of supplies and Logistics.

To continue diagnosing the business on a yearly basis.

Implement the management of production orders and production scheduling through software and appropriate hardware.

Implement the strategic management and capacity through the implementation of customized software to be set.

As already selected, the 20 weaknesses of practices and performance according to the correlation (ri), taking into account all the indicators evaluated, it is understood that with this list of indicators is possible to prepare action plans (PDCA) and/or projects improvements and thus better structure the company's strategic plan implementations to improve 8% in the evaluation of general indicators of Practice to promote the classification in challenging category.

The EDM, which is a pushed production programming system, concludes that there is discomfort about the problems with suppliers, resulting in daily actions to "put out fires" and long lead times.

The stock waste in the process is visible and is an opportunity for improvement, according to the evaluations of managers. Investments in machinery happen, but you need to use existing resources, sample software. We suggest the continuity of staff training in investments in order to add value to the customer, aligned with the strategic objectives for achieving greater financial results.

The team shows ability to acquire knowledge and skill as it has the maturity and attitude to explain critically about their problems, the environment is favorable for transforming problems into opportunities for financial gain.

The company has the opportunity to achieve greater results with the routine management of the day-to-day management and the capacity of the equipment, but you need Hoshin planning to guide all levels of integrated, participatory and motivated by the change of culture , turning to the continuous appreciation of the people, delegation of responsibilities (Jidoka), for the improvement of quality and productivity and generate sustainable results for business.

6. EDM RESULTS IN DIFFERENT COMPANIES

The Lean Enterprise approach was structured through five basic principles that can meet many business segments. They are: Producing only on demand; Robust product; Challenge and encourage employees; Prevent defects and Avoid waste. These principles were generated by research and applications over the years by the investigator and based on the guidelines of a leading European household appliance market.

It is a consensus in the surveyed companies that Lean transformation based on tools, presented methods and principles can and should be aligned at all levels of the organization and integrated and strategic way to facilitate the achievement of business excellence. It was understood that it goes through a process of social technical improvement that involves everyone and it takes years to achieve Lean Transformation. Research and knowledge of tools and Lean methods were essential to the realization of the EDM of the workshop in the surveyed companies.

In all the four researched companies , there was an intention to implement a strategic management system and move to implement the EDM annually to improve the measurement of the performance integrated and aligned with the Hoshin planning and managing the indicators frequently in the future.

6.1 EDM Indicator Case Analysis

It was found in the company's rating according to the analogy to the boxing match of benchmarking that the real opportunity is to improve the weak points, i.e. where the company is further from the excellence. Each company has its operational performance and evaluates the areas of diagnosis, it is noted that there are differences by company and also by area. By analyzing the table 13 and figure 16 we can see where there is a need to realize projects and improvements in management.

Table 14 Results of EDM indicators in the surveyed companies

Diagnosis industrial areas (%)	Machining	Plastic	Textile	Sanitation
Total quality PR	45%	43%	53%	45%
Total quality PF	83%	43%	68%	55%
Lean Production PR	53%	36%	58%	42%
Lean Production PF	60%	42%	56%	69%
Logistics PR	45%	40%	65%	60%
Logistics PF	69%	40%	63%	74%
Organization and Culture PR	52%	32%	70%	48%
Organization and Culture PF	40%	40%	80%	40%
New Product Performance PR	62%	62%	70%	40%
New Product Performance PF	67%	30%	77%	43%
Innovation management PR	33%	40%	70%	47%
Innovation management PF	70%	50%	70%	20%
Environment, Health and Security PR	65%	15%	78%	53%
Environment, Health and Security PF	70%	25%	75%	55%
General Performance Indicators	80%	60%	80%	80%
General Practice Index PR	52%	39%	66%	46%
General Performance Index PF	69%	39%	67%	59%

The machining company is in the category "Vulnerable", as obtained in the general indicator of Practice 52% and overall indicator Performance 69%. The research company will strive to achieve classification in challenging category, so the general indicator of Practice will get an improvement of 8% in the evaluation, to move from 52% to 60% in the general indicators of Practice.

The plastic parts injection company is in the "punching bag" category, obtained in the general indicator of Practice a score of 39% and overall performance indicator of 39%. The researched company's challenge will be to achieve the classification in the "Challengers", so the general indicator of practice and performance will need to get an improvement of 21% in the evaluation, to rise from 39% to 60% in the general indicators of Practice and performance.

The textile company lies in the "Challenger" category which obtained 66% in the general indicator of Practice and overall performance indicator of 67%. The researched company's challenge will be to achieve the classification in world class manufacturing category, so the general indicators of practice and performance will need an improvement of around 14% in the evaluation, to spend 66% on average for 80% of the indicators general Practice and Performance.

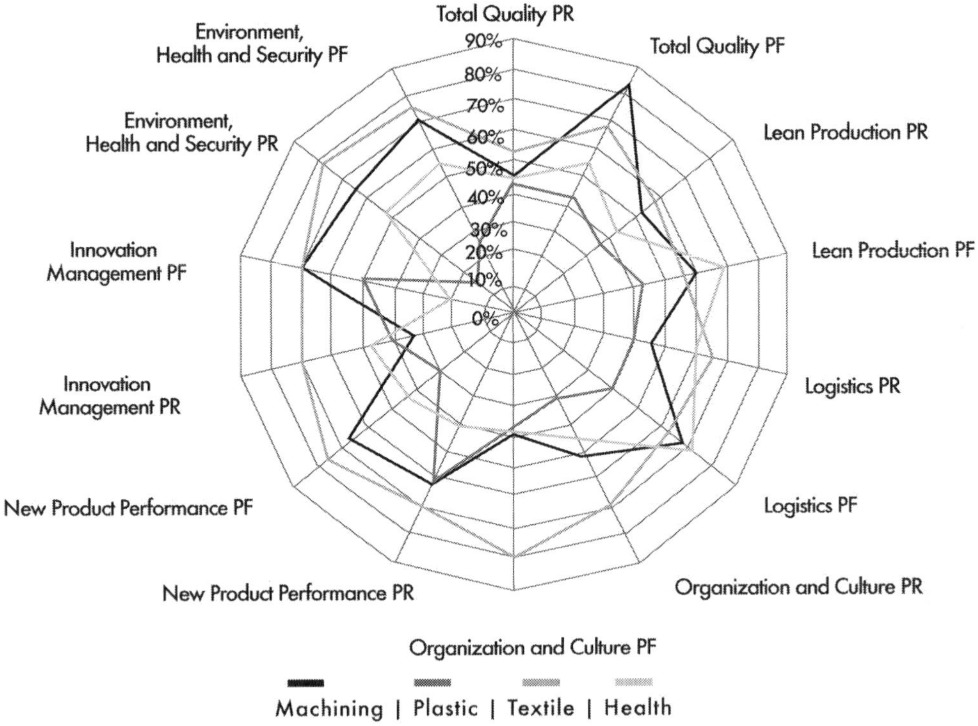

Graph 16 Comparison of the four cases studied

The Basic Sanitation company is in the "counterweights" category which obtained in the general indicator of Practice a score of 46% and overall performance indicator of 59%. The researched company's challenge will be to achieve the classification in the "Challengers" category, so they will have to obtain an overall indicator of 14% improvement in the assessment, and from 46% to 60% in general practice indicators.

The four companies are far from being considered World Class level, which correspond to values above 80%, companies compared to other segments did not have a good performance (Figure 53).

According to the research conducted throughout this work the use of cost-cutting strategy coupled with the growth strategy, identification and formalization of the company's principles and the implementation of strategic planning and strategic management are recommended.

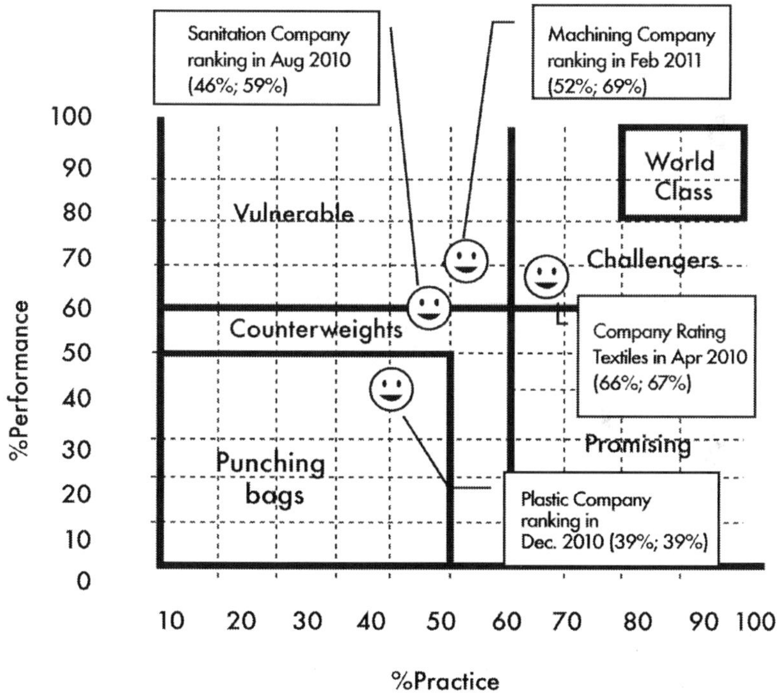

Figure 53 Classification of four companies surveyed

Historical information was obtained on the industrial benchmarking database (IEL), and the comparison between the four companies surveyed and the average of the worldwide industry leaders and three similar segments (electronics companies, plastic and textile) is conducted. Although it can be used to compare companies in the same segment, in the case of EDM the main focus of comparing companies is not the same but the point is to identify of the ranking, the selection and evaluation of indicators, i.e., the identification of strengths and weaknesses to identify project requirements and then implement a good hoshin planning, see Figure 54.

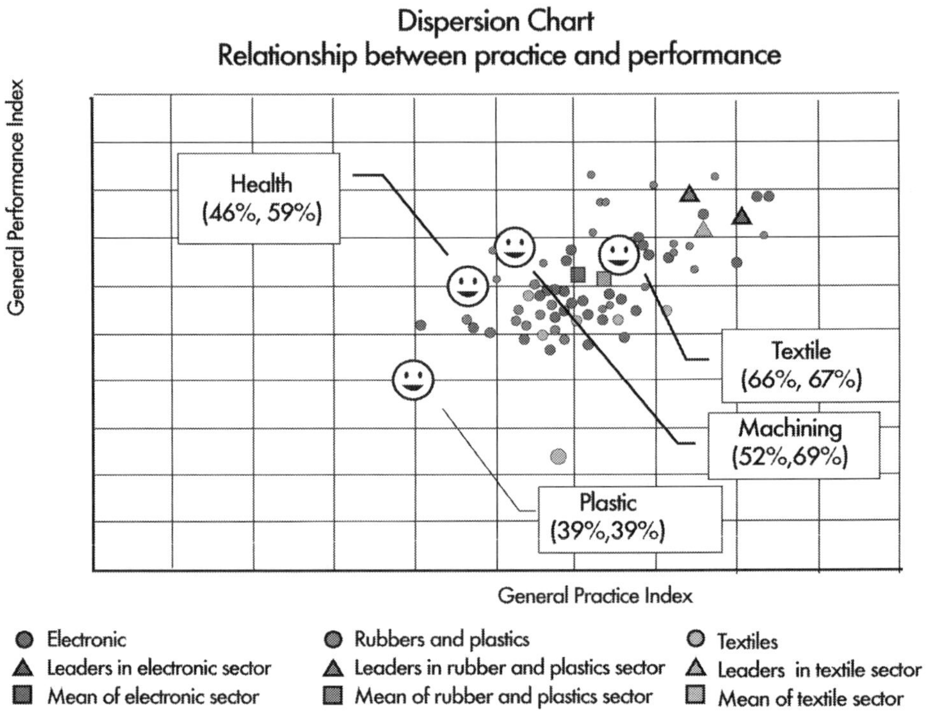

Figure 54 Classification of three different segments and the four companies surveyed (adapted from IEL / CTI, 2010)

Faced with the possibility to improve the score of the diagnosed companies by reducing waste and adding value to customers, you can use the principles, methods and tools of Lean Enterprise. There will be a need to revise the objectives and strategies for a sustainable business in order to define some strategic planning, therefore, investment projects for the improvement and achievement of results should be drawn according to the business strategy and not pushed empirically.

In the fifteenth to seventeenth stages of EDM, it is possible to create action plans to be implemented, with the identification of those responsible for implementation and deadlines. The researcher, as well as trained professionals are able to moderate and help create action plans to explore opportunities for improvement identified in comparison with leading companies. In this research action plans were not performed, because according to strategic management model this is an activity that belongs to the post strategic planning application of EDM.

While there may be the need to adapt the EDM for service companies, it is recommended to the companies surveyed, with conditions practices and performances and similar cultures to include a set of basic actions to increase assets and strategic planning. The actions are:

Implement cost reduction strategy coupled with the growth strategy through workshops with management team.

Implement the Hoshin planning and strategic management, set goals and objectives for all areas of integrated and transparent way, if possible with software support.

Implement the management of routine day-to-day, starting with the work pattern of critical activities and processing bottlenecks.

To systematize a method of Company troubleshooting.

Implement project 5S and visual management in processes.

Develop the mapping of the current and future value stream to guide improvements.

Implement the autonomous maintenance, leverage the TPM, to reduce dwell times and involve operational staff.

Develop a matrix of multi-functionality.

Leverage a plan of ideas with direct awards and leverage the career path.

Develop training plan for managers and employees to implement the Lean Enterprise.

Leverage cost reduction and partnerships in supply management and Logistics.

To continue the business diagnosed annually with six-monthly review.

Implement the management of production orders and production sequencing through appropriate software and hardware.

Implement the strategic management and capacity through the implementation of customized software.

The four companies surveyed have the opportunity and the best recommendations to manage their capacity to gain better results for the organization. We consider EDM to be a very applicable tool to use in business and can contribute to the sustainability of enterprises.

6.2 Analysis of results according to statistical tools

For now this research was carried out by classifying according to the boxing analogy to identify how the four companies surveyed are far from being world class, the dealings of the data are in percentage and is considered a semi-qualitative evaluation. With the proposal to have a closer decision parameter of quantitative result it was decided to use the boxplot tool among other means.

We present the results from the statistical perspective using Minitab software, widely used in Six Sigma approach and in engineering, this explores the results through three different graphics: histogram, boxplot and confidence interval, always comparing the four companies surveyed. Note that you can also analyze the areas of each company and compare them according to the graphs.

Comments on Grey Calculations

With the Grey (r) analysis correlation, you can see which indicator has the highest value, that is, the best performance according to the evaluation of individuals and the lowest value indicator because correlate the grades assigned. Later on, once informed of the low performance indicators, one can analyze the root cause of problems and then develop a recommendation plan to enhance a particular performance issue. of . The analysis and recommendations for improvements are carried out according to the reviews and record (in a spreadsheet) of the evaluators themselves.

All the evaluators first assign a score of one to five for each indicator, and as the theory of Grey system assigns the weight for all evaluators, starts from the principle in the EDM that everyone has the same degree of importance with respect to its opinion, it is assumed in this document that each person is a referee of equal importance in the business process.

Over 80% is world class according to boxing analogy benchmarking. Benchmarking predicts that data reaches 80%. When looking at the textile company there is a better result when it comes to percentage, it reached 66% and 67% of performance practice, the textile company when looking at percentage it appears that she is better than the other three companies, i.e. is closest to being a world-class company (see Figure 54 - Classification of four companies surveyed)

The calculation basis for defining the percentages of performances and practices is in a spreadsheet. In the case of textile company in Figure 54 classification of the four companies surveyed show the result of the notes in percentages of 66% in performance and 67% in practice. This percentage is the result of notes 1-5, given by the evaluators, with the note an expectation, semi-quantitative assessment.

Later the calculation of Grey correlation analysis is applied with the same notes, in which the sample application included in the technical report example (CALADO, BATOCCHIO, CALARGE, 2011a) and have been described step by step in Chapter 3, based on item Tie-jun and Sha (2008).

The highest value is a variable, is the largest note that the evaluator attributed. If the percentage is a part for the whole and in the case of matrix x for Grey system this is not percentage. To define the percentage of performance and practice is considered Note 5 as the maximum value, so this mark is equivalent to 100%.

The Grey system is a quotient, where the relation is not based upon how much each evaluator provides a grade compared to a max. assigned by the same evaluator. The Grey system is a relationship with the maximum per each evaluator and not a relationship with the general maximum.

As the Grey system does not have a constant denominator in part, the Grey system has maximum and the maximum is not necessarily the whole, undivided by a single constant value as is the case of the percentage of the performance calculation.. Thus, the system is called the Grey correlation, which is calculated as r integer ranging from 0 to 1, being different from the percentage calculation. In the matrix x of the Grey calculation, when the highest score of the evaluators is assigned equal to 4 (highest grade given), this value will match the maximum value that is equal to 1 and has nothing to do with percentage and are different from the percentage of benchmarking.

The matrix x indicates all levels from 0 to 1, making a ratio between the value observed at the higher the same evaluator. When observing all the notes reviewer the highest score will be equivalent to index r = 1.

In the case of the Grey system, equal weights were assigned considering the same importance to all evaluators, but you can assign different weights according to the evaluator or project in question.

In the continuation of calculation, ρ is assumed constant at 0.3, and as Tie-Jun Sha (2008), which is assigned to mention that from 0.1 to 0.5, as the authors investigated assign a value for ρ called as the distinguishing factor for Deng (1982), having performed in this research simulations with different values ρ ($\rho = 0.1$ and $\rho = 0.5$), but it does not bringing big difference to the results as of ρ assumed value.

Comments on application of boxplot

Considering the box plot, the maximum and minimum traces are increasingly disparate, more asymmetric, the more the average increases. By boxplot (Graph 17) can be compared to 4 companies through quartile who is the low trace of the box, two who quartile is the median, quartile three that is the top feature of the box, minimum and maximum values . The outlier is a peak that exceeds the upper limit for the given data according to the theory of Boxplot.

It is observed that in the boxplot quartile, median, mean increased compared with one another and that issome sort of evidence of the exploratory analysis to point out a specific level that suggests that the values of machining company, in this case, tend to be larger than the plastic company, cleaning and textile industries. The Graph shows in all examples that there are major problems in all four cases due to the small values. The main problems can be found in the plastic company. As for the machining company the problems are related to leveling.

On the other hand there is the question of variability, it means that some values here have reached the level of better performance, but are rare or few values. In the case of plastic company or outliers reach a value close to good performance. Only the textile and machining company achieve better performance values, but still only reach the discrepancy, few values, that is, one or the other what is causing it.

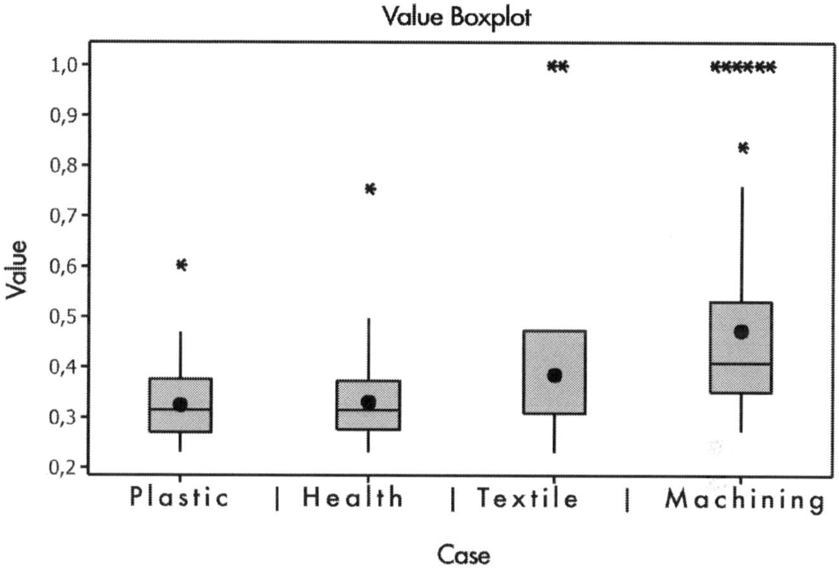

Graph 17 Boxplot of the 1st quartile of the 4 cases studied

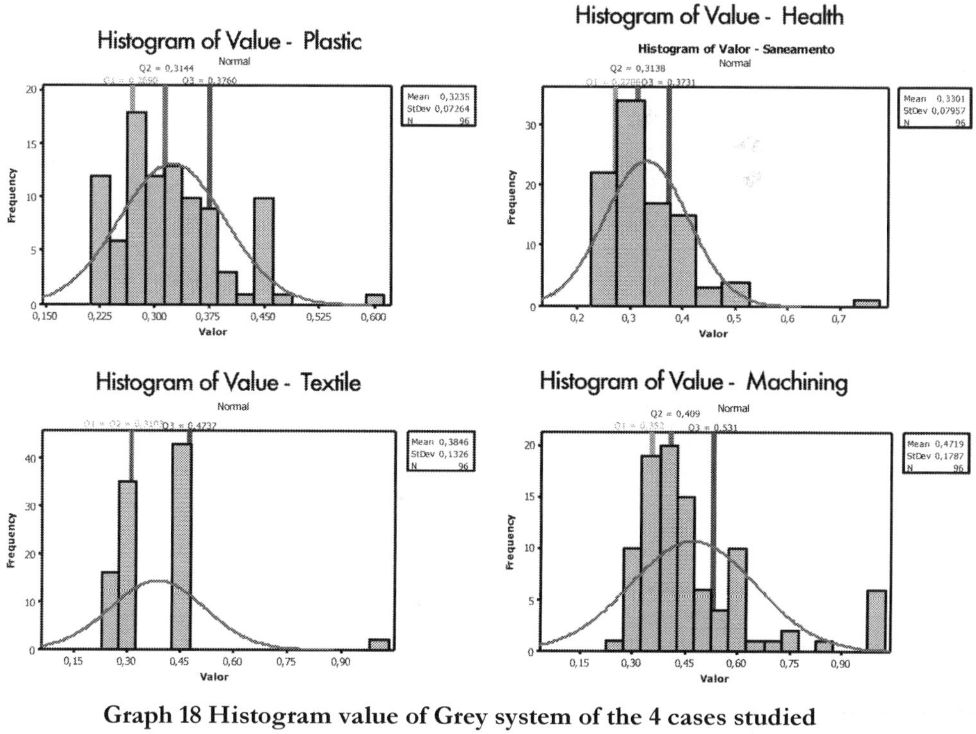

Graph 18 Histogram value of Grey system of the 4 cases studied

For instance the lower the level of rental data shows most symmetric distribution of the data, that is, the distribution is very close to the average median.

The plastic company's lease is with the median value Q2 = 0.3144 and your grade average of 0.3235 which means little distance between median and average. When there is little distance means that the data is not so this asymmetry circumstances compared to other cases, for example, the sanitation company has a greater distance between mean and median.

Comments on application of the confidence interval

The range chart (Graph 19) shows that with increasing lease, as the normal curve will move towards higher values on average, while it also ends up having an increased variability, i.e. the plastic and sanitation has a similar behavior.

It is proved by the confidence interval that the difference there is significant and it is no coincidence, the graph has an average estimate for each company represented by points.

The Box plot is different from the confidence interval, both have different opinions, each tool refers to another side of statistics, can now be said that it is no coincidence that the means are different. The sample is possible to know which companies are best and with evidence.

What matters now is to have an estimated sample mean, and the range can be assured with confidence margin. The average error which has to reliability of 95% is within the range, may be noted that the range in machining company is much larger than the range of the plastic company. Plastic company data are more homogeneous and smaller also.

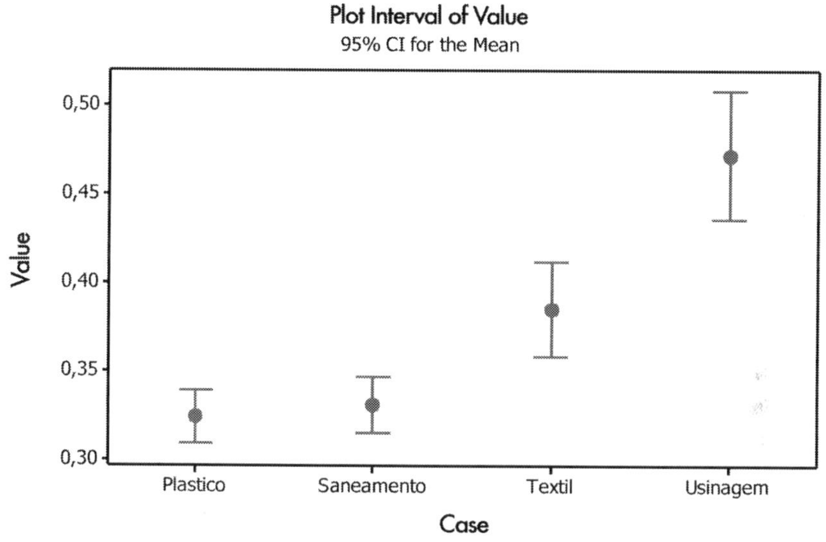

Graph 19 Confidence interval of 4 cases investigated

Now with the use of the confidence interval chart, it is seen that all companies are bad, which means a low level of performances and practices, which has a guarantee of a 95% chance that the true average is in the range, has a significance of 5% margin of error (statistical inference) with 2.5% above and below 2.5%.

It is also observed that between the plastic company and sanitation there is a technical tie, the trends are similar because they are at the same height, so the boxes are the same height.

To show the difference between the companies the boxplot chart can be used to see the data distribution and the confidence interval chart to see the distribution of the means of data.

You can be sure of the difference between companies because on average the machining company is significantly higher than the textile company, and in turn is significantly sanitation company and which in turn has a technical tie with the plastic company i.e. the sanitation and plastics companies are not significantly different, even though differences between means, the means are included in the same range, the ranges of the two companies cover the means of the same.

The graphs are distant because of the asymmetry. The Boxplot and the confidence interval helps to see the best performance. The only confidence interval of the problem to the data, when using this tool for decision making is that the average shifts up because of outliers.

6.3 Final thoughts

The companies studied in this investigation are far from being world class enterprises, it is necessary to improve performances and practices, evaluate and compares the correlation of company performance indicators is confirmed that all four companies in this sample are low maturity . In all companies surveyed we find specific actions, which isolated as a rule, are not sustainable.

A key finding of this research is the development and application of the Enterprise Diagnostic Method, in which we presents a detailed roadmap for a better understanding and application. The EDM creates the conditions and assists in the medium and long term decisions, allowing you to plan actions against the Hoshin planning and strategic management more integrated and sustainable manner. In summary is presented below the main contributions of this book:

The Enterprise Diagnostic Method presented in this book has structured guidelines according to the PDCA, through the use of a set of concepts relevant to the management and sustainability of the business, such as: Lean, Benchmarking, correlation system and Grey analysis; Hoshin planning (Honshin Kanri), strategic management and others.

The Lean Enterprise approach was structured within this work through five basic principles that can meet many business segments. It was understood that it goes through a process of social improvement in the technical "top-down" perspective, which involves all people and it takes years for the Lean transformation.

The definitions and knowledge of Lean tools and methods were essential to the realization of the EDM workshop at the companies studied. Warning that the result of applying Enterprise Diagnostics Method in companies depend in part on the knowledge, skill and attitude of the researcher (the producer of the business diagnose) and direction and the proper use of action research methodology (for specific knowledge and lead the work team learning).

The EDM creates the conditions and assists in the medium and long term decisions, allowing you to plan actions against the Hoshin planning and strategic management more integrated and sustainable way to transform a conventional production Lean system.

It is noteworthy that the implementation of action research, with this strategy along with the various groups of managers and specialists in the surveyed companies that attended the workshops for several weeks have brought results ranging from knowledge in the workplace to the gradual change in behavior and culture.

In this sample of companies, from the management side, there was an intention to implement a strategic management system and to implement the EDM annually to enhance the measure of performance, integrated and aligned with strategic planning and management of indicators frequent way.

The identification of strengths and weaknesses to define projects and conducted with the participation of all levels (leaders of the companies) can provide aid subsidies for investment decisions, strategic information and opportunities to improve processes by transforming the Lean approach.

The EDM changes the conventional mental model and becomes a semi-quantitative assessment of the degree of organizational maturity of enterprises due to the use of a set of interrelated knowledge and suggests a customized way to implement self-assessment, internal benchmarking and deploy targets and guidelines in companies. With the application of EDM, managers are involved and have a better understanding of the situation of your business as a whole, then define its priority projects in agreement and the definition of what tools, methods and optimization approaches and improving processes become an aligned to the strategic objectives defined decision to strategically manage.

APPENDIX A - MANAGEMENT MODEL HOSHIN
(Jackson, 2006)

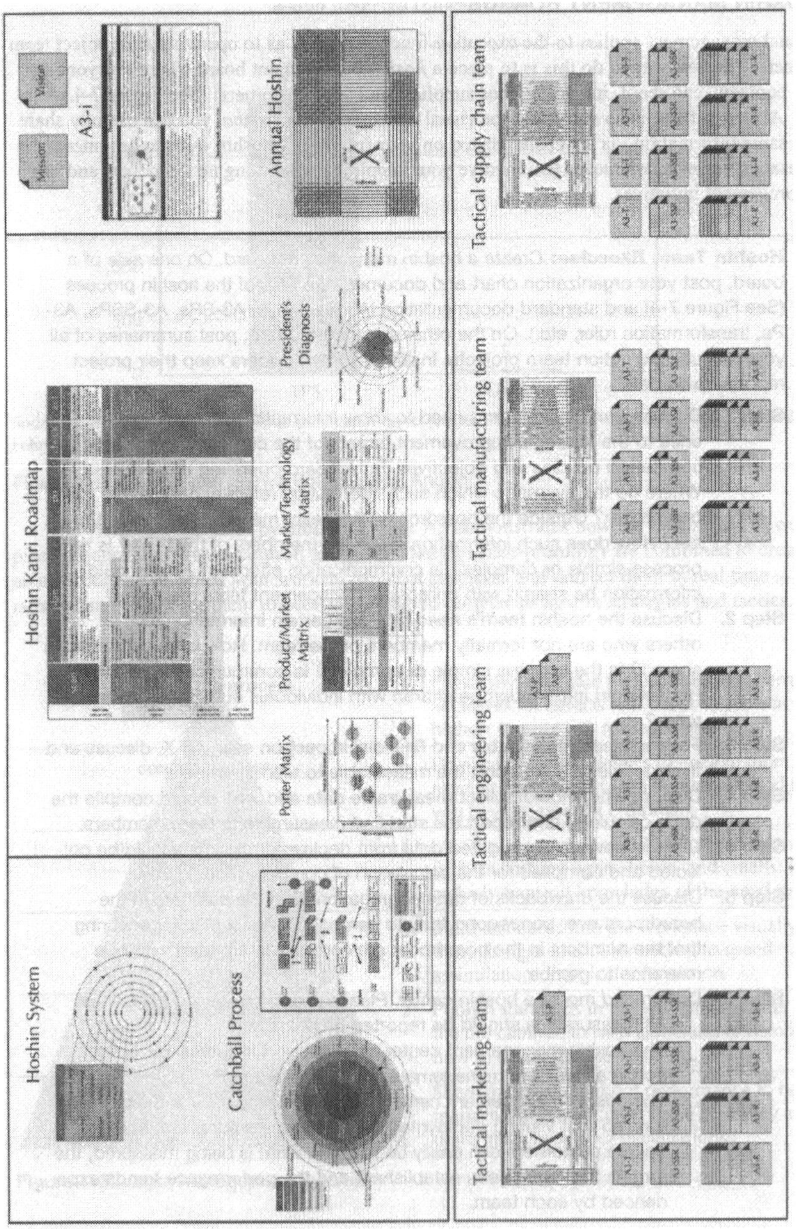

CONSULTED BIBLIOGRAPHY

ABDULMALEK, F. A., RAJGOPAL, J. Analyzing the benefits of lean manufacturing and value stream mapping via simulation: A process sector case study. Int. J. Production Economics, Volume 107, 2007, p. 223-236.

AKAO, Y. Desdobramento das diretrizes para o sucesso do TQM. Porto Alegre: Bookman, 1997.

ALVES, R., CALADO, R. D. Metodologia de estudo de tempos para aumento de produtividade em posto de trabalho: utilizando a técnica de MTM. In: Simpósio Internacional de Engenharia Automotiva - SIMEA 2005, 2005, São Carlos.

ALVES, R., OLIVEIRA, T. H., CALADO, R. D. Balanceamento de Linha: análise de agregação de valor para o desenvolvimento de método de trabalho. In: CCM 2007 - Congresso Corte & Conformação de Metais, 2007, São Paulo.

ALVES, R., OLIVEIRA, T. H., CALADO, R. D., CALARGE, F. A. Aplicação da Técnica SMED e Metodologia DMAIC: experiência no segmento de eletrodomésticos -linha branca. In: CCM 2007 - Congresso Corte & Conformação de Metais 2007, 2007, São Paulo.

ANUÁRIO ESTATÍSTICO, 2012, Indicadores industriais, consultado em 16/11/14 e disponível no site: http://www.mdic.gov.br/arquivos/dwnl_1337260114.pdf

BALLOU, R. H. Gerenciamento da Cadeia de Suprimentos / Logistics Empresarial. Porto Alegre: Bookman, 2006.

BRAZ, M. A. Ferramentas e gráficos básicos. In ROTONDARO,R. G.(coord.) Seis Sigma - Estratégia Gerencial para a Melhoria de Processos, Produtos e Serviços. São Paulo: Atlas, 2008.

CALADO, R. D. Aplicação de Conceitos da Manufatura Enxuta no Processo de Injeção e Tampografia de Peças Plásticas. Dissertação de mestrado acadêmico apresentada à comissão de Pós Graduação da Faculdade de Engenharia Mecânica, como requisito para a obtenção do título de Mestre em Engenharia Mecânica. Unicamp, Campinas, SP, 2006.

CALADO, R. D., BATOCCHIO, A., CALARGE, F. A. Relatório da aplicação do Método de Diagnóstico de Empresa (BDM): uma aplicação no segmento metal mecânico – usinagem de precisão. Relatório técnico, Campinas, ISBN 978-85-912489-3-3, 2011a.

CALADO, R. D., BATOCCHIO, A., CALARGE, F. A. Relatório da aplicação do Método de Diagnóstico de Empresa (BDM): uma aplicação no segmento plástico – injeção de peças de plástico. Relatório técnico, Campinas, ISBN 978-85-912489-0-2, 2011b.

CALADO, R. D., BATOCCHIO, A., CALARGE, F. A. Relatório da aplicação do Método de Diagnóstico de Empresa (BDM): uma aplicação no segmento de serviço público – saneamento básico. Relatório técnico, Campinas, ISBN 978-85-912489-1-9, 2011c.

CALADO, R. D., BATOCCHIO, A., CALARGE, F. A. Relatório da aplicação do Método de Diagnóstico de Empresa (BDM): uma aplicação no segmento têxtil – embalagem industrial. Relatório técnico, Campinas, ISBN 979-85-912489-2-6, 2011d.

CALADO, R. D. ; BATOCCHIO, A. ; CALARGE, F. A. . Diagnóstico Industrial: melhorias do gerenciamento da capacidade e otimização dos processos. In: XVIII Simpósio Internacional de Engenharia Automotiva - SIMEA 2010, 2010, São Paulo. XVIII Simpósio Internacional de Engenharia Automotiva - SIMEA 2010, 2010.

CALADO, R. D., BIONDI, A. Aplicação da metodologia MPT e OEE: exemplo de boas práticas para aumentar a disponibilidade de prensa. In: CCM 2007 - Congresso Corte & Conformação de Metais, 2007, São Paulo.

CALADO, R. D. CIRILLO, C.C., CARVALHO, M. F. A Metodologia Kanban: os benefícios gerados para um fabricante de eletrodomésticos e seus fornecedores via internet. In: III Congresso Brasileiro de Engenharia de Fabricação, Joinville, 2005.

CALADO, R. D., CALARGE, F. A. Gerenciamento da Produção através do OEE: caso de uma empresa de eletromecânica. In: CCM 2009 - Congresso Corte & Conformação de Metais, 2009a, São Paulo.

CALADO, R. D., CALARGE, F. A. Conceitos do Projeto Axiomático e desdobramento de Metas: aplicação em uma empresa do setor de serviços In: SIMPOI 2009 - XII Simpósio de Administração da Produção, Logistics e Operações, 2009b, São Paulo.

CALADO, R. D., LIMA, P. C. Aumento da capacidade dos Equipamento com a utilização da Metodologia OEE. In: Congresso de Corte e Conformação de Metais, 2005, São Paulo.

CALADO, R. D., LIMA, P. C. A Aplicação da metodologia de Célula de manufatura na Solução de Risco de Acidente: uma experiência em um fabricante de eletrodomésticos. In: II Congresso Brasileiro de Engenharia de Fabricação, São Paulo, 2003,

CALADO, R. D., LIMA, P. C., CALARGE, F. A., MULLER, R. A. Metodologia de estudos de tempos e balanceamento de linha de montagem: utilizando a técnica de filmagem. In: III Congresso Brasileiro de Engenharia de Fabricação, Joinville, 2005.

CALADO, R. D., RUGGIERO, S., COOPER, R. O Mapeamento da Cadeia de Valor na Transformação de Chapas. In: II Congresso de Corte e Conformação de Metais, São Paulo, 2003.

CALARGE, F. A., CALADO, R. D. A Metodologia de troca rápida de ferramentas aplicadas na conformação: a experiência de um fabricante nacional de eletrodomésticos da linha branca. In: Congresso Corte & Conformação de Metais, São Paulo, 2001.

CAMP, R.C. Benchmarking: The Search for Industry Best Practices that Lead to Superior Performance. ASQC Quality Press, Milwaukee, WI, 1989.

CAMPOS, V. F. Controle da Total quality (no estilo japonês). Nova Lima: INDG, 2004a.

CAMPOS, V. F. Gerenciamento da rotina do trabalho do dia-a-dia. Nova Lima: INDG, 2004b.

CAMPOS, V.F. Gerenciamento pelas diretrizes (Hoshin Kanri). Nova Lima: INDG, 2004c.

CARVALHO, M.M., PALADINI, E. P. Gestão da qualidade: teoria e casos. Rio de Janeiro: Elsevier, 2005.

CERTO, S.C., PETER, J.P. Administração estratégica: planejamento e implantação da estratégia. São Paulo: Pearson Education do Brasil, 2010.

CHAN, J. W. K., TONG, T. K. L. Multi-criteria material selections and end-of-life product strategy: Grey relational analisys approach. Material and Design, v 28, 2007, p. 1539-1546.

CHAVES, N. M. D. Caderno de campo das equipes de melhoria contínua. Nova Lima: INDG, 2006.

CHOPRA, S.; MEINDL, P. Gerenciamento da Cadeia de Suprimentos: Estratégia, Planejamento e Operação. São Paulo: Pearson Prentice, 2003.

CHRISTOPHER, M. Logistics e Gerenciamento da Cadeia de Suprimentos: criando redes que agregam valor. São Paulo: Thomson Learning, 2007.

CORRÊA, H. L. Gestão de redes de suprimentos: integrando cadeias de suprimentos no mundo globalizado. São Paulo: Atlas, 2010

CORRÊA, H. L.; CORRÊA,C.A. Administração de Produção e Operações. São Paulo: Atlas, 2007.

CUDNEY, E. A. Using Hoshin Kanri to Improve the Value Stream. New York: Productivity, 2009.

DAILEY, K. W. The Lean Manufacturing Pocket Handbook. USA: DW Publishing Co., 2003;

DAUGHERTY, P. J., PITTMAN, P.H. Utilization of time-based strategies Creating distribution flexibility/responsiveness. International Journal of Operations & Production Management, Vol. 15, No. 2, 1995, pp. 54-60.

DENG, J. L. Control problems of Grey Systems. Systems & Control Letters, Vol.1, Issue 5, March, 1982, p.288-294.

DENG, J. L. Introduction of Grey System. Journal Grey Systems, v 1, nº 1, 1989, p.1-24.

DENNIS, P. Getting the right things done: a leader's guide to planning and execution. Cambrige, MA, USA: Lean Enterprise Institute, 2006.

DENNIS, P. Produção Lean Simplificada. Porto Alegre: Bookman, 2008.

EATON, Lean Manufacturing: Guia de Bolso da Eaton Lean Six Sigma - ELSS. Valinhos: 2006.

FELD, W.M. Lean manufacturing: tools, techniques, and how to use them. CRC, 2000.

FERREIRA, A.B.H. Novo dicionário da língua portuguesa. São Paulo: Positivo, 4ª ed., 2009.

GHINATO, P.. Sistema Toyota de Produção: Mais do que Simplesmente Just In Time. 1.ed. Belo Horizonte: Revista produção, v.5, n.2, p. 169-189, nov. 1995.

HANSEN, R C. Eficiência Global dos Equipamentos: uma poderosa ferramenta de produção / manutenção para o aumento dos lucros. Porto Alegre: Bookman, 2006.

HANSON, P.; VOSS, C. A. Benchmarking Best Practice in European Manufacturing Sites. Business Process Re-engineering&Management Journal, v. 1, n. 1, p. 60-74, 1995.

HOLWEG, M. The genealogy of lean production. Journal of Operations management. vol.25, pp.420-437, 2007.

IEL/SC - INSTITUTO EUVALDO LODI DE SANTA CATARINA. Banco de dados do Benchmarking Industrial. Florianópolis, 2011

IEL/SC - INSTITUTO EUVALDO LODI DE SANTA CATARINA. Manual de aplicação do bencmarking industrial. 2006

JACKSON, T. L. Hoshin Kanri for the Lean Enterprise: developing competitive capabilities and managing profit. New York: Productivity, 2006.

JACOBS, F. R. Administração da Produção e de Operações: o essencial. São Paulo: Saraiva, 2009.

JURAN, J.M. A qualidade desde o projeto: os novos passos para o planejamento da qualidade em produtos e serviços. São Paulo: Cengage Learning, 2009.

KAPLAN, R. S.; NORTON, D. P. A estratégia em ação: balanced scorecard. Rio de Janeiro: Elsevier, 1997.

KONDO, Y. Hoshin Kanri - a participative way of quality management in Japan, The TQM Magazine, Vol. 10 No. 6, 1998, pp. 425-31.

KRAJEWSKI, L; RITZMAN, L.; MALHOTRA, M. Administração de produção e operações. 8 ed. São Paulo: Pearson Prentice Hall, 2009.

LIKER, J. K.; HOSEUS, M.. A cultura Toyota: o modelo Toyota aplicado ao desenvolvimento de pessoas. Porto Alegre: Bookman, 2009.

LIKER, J. K.; Meier, D. O Modelo Toyota: manual de aplicação. Porto Alegre: Bookman, 2007.

LIKER, J. K.; Meier, D. O Talento Toyota: o modelo Toyota aplicado ao desenvolvimento de pessoas. Porto Alegre: Bookman, 2008.

MIYAKE, D. I. Melhorando o processo: seis sigma e sistemas de produção Lean. In ROTONDARO,R. G.(coord.) Seis Sigma - Estratégia Gerencial para a Melhoria de Processos, Produtos e Serviços. São Paulo: Atlas, 2008.

MORGAN, J. M., LIKER, J. K. Sistema Toyota de desenvolvimento de produto: integrando pessoas, processo e tecnologia. Porto Alegre: Bookman, 2008

MURMAN, E., McMANUS, H., CANDIDO, J. LAI Lean Academy. Participant Book, PI.23s. Cambridge: Lean Advancement Initiative, July 19-21 2010.

MURMAN, E.; et al. Lean Enterprise Value: insights from MIT's Lean Aerospace. New York: Palgrave Macmillan, 2002.

NAKASHIMA, S. Introduction to TPM: Total Productive Maintenance. Cambridge, MA: Productivity, 1988.

NOVAES, A. G. Logistics e Gerenciamento da Cadeia de Suprimentos. Rio de Janeiro: Elsevier, 2007.

OHNO, T. O sistema Toyota de Produção: além da produção em larga escala. Porto Alegre: Bookman, 1997.

REICHHART, A., HOLWEG, M. Creating the customer-responsive supply chain: a reconciliation of concepts. International Journal of Operations & Production Management. Vol. 27, No. 11, 2007, pp. 1144-1172.

ROTHER, M.; SHOOK, J. Aprendendo a enxergar: mapeando o fluxo de valor para agregar vlor e eliminar desperdício. São Paulo: Lean Institute Brasil, 1998.

SAATY, T. L. The analytic hierarchy process: planning, priority setting. New York: McGraw Hill International, 1980.

SAMOHYL, R. W. Controle Estatístico de Processo e ferramentas da Qualidade. In CARVALHO, M.M.; PALADINI, E.P. (coord.) Gestão da Qualidade: teoria e casos. Rio de Janeiro: Elsevier, 2005.

SATOLO, E..G., CALARGE, F. C.; SALLES, J. A. A.; MAESTRELLI, N. C.; PAPA,

M. C. O.; ABACKERLI, A. J. Uma análise sobre questões atuais do Sistema Lean Production: um estudo exploratório de um site internacional de discussões. In: Simpósio Internacional de Engenharia Automotiva, 14, São Paulo, 2006.

SCHUCH, C.G.; CASTRO, V.F. Uma breve análise sobre a competitividade das empresas líderes no mundo, sob a ótica do Benchmarking Industrial. Disponível em: http://www.ielsc.org.br/web/pt/info/artigo-benchmarking-industrial# , acesso em: 18/03/2011.

SEIBEL, S. Um Modelo de Benchmarking baseado no Sistema Produtivo Classe Mundial para Avaliação de Práticas e Performances da Indústria Exportadora. 173f. Tese (Doutorado). Programa de Pós-graduação em Engenharia de Produção. Universidade Federal de Santa Catarina. Florianópolis, 2004.

SEIBEL, S. et al. Benchmarking: a busca pela classe mundial. In: CONGRESSO BRASILEIRO DE ENGENHARIA DE FABRICAÇÃO, Anais...Curitiba/PR, 2001.

SHINGO, S. Fundamental Principles of Lean Manufacturing. Bellingham: Enna Products Corporation, 2009.

SHINGO, S. O Sistema Toyota de Produção do Ponto de Vista da Engenharia de Produção. Porto Alegre: Bookman, 1996a.

SHINGO, S. Sistema de Produção com Estoque Zero: o sistema Shingo para melhorias contínuas. Porto Alegre: Bookman, 1996b.

SHINGO, S. Sistema de Troca Rápida de Ferramentas: uma revolução nos sistemas produtivos. Porto Alegre: Bookman, 2000.

SIMCHI-LEVI, D., et al. Cadeia de Suprimentos: Projeto e Gestão. 3ª ed., Porto Alegre: Bookman, 2010.

SLACK, N. , et al. Gerenciamento de Operações e de processos: princípios e práticas de impacto estratégico. Porto Alegre: Bookman, 2008.

SLACK, N., LEWIS, M. Estratégia de Operações, Porto alegre, Bookman, 2009, p.135.

SLACK, N. et al.. Administração da Produção. 3. ed., São Paulo: Atlas, 2009

SOBEK II, D. K.; SMALLEY, A. Entendendo o pensamento A3: um componente crítico do PDCA da Toyota. Porto Alegre: Bookman, 2010.

TAYLOR, D. A. Logistics e Gestão da cadeia de suprimentos: uma perspectiva gerencial. São Paulo: Pearson Addison-wesley, 2005, p.87.

TIE-JUN, C.; SHA. L. Application and Study of Lean Production Theory in the Manufacturing Enterprise. IEEE 2008, 2008 International Conference on Information Management, Innovation Management and Industrial Engineering. 2008.

TRIOLA, M. F. Introdução à estatística. 10ª ed., Rio de Janeiro: LTC, 2008.

VOSS, C. A.; AHLSTRÖM, P.; BLACKMON, K. Benchmarking and operational performance: some empirical results. International Journal of Operations & Production Management, Vol. 17 No. 10, 1997, pp. 1046-1058.

WEE, H.M.; WU, S. Lean supply chain and its effect on product cost and quality: a case study on Ford Motor Company. Supply Chain Management: An International Journal, Volume 14, N. 5, 2009, p.335-341.

WITCHER, B.J. Hoshin Kanri: a study of practice in the UK, Managerial Auditing Journal, Vol. 17 No. 7, 2002, pp. 390-396.

WOMACK, J. P.; JONES, D. T., ROOS, D. A máquina que mudou o mundo: baseado no estudo do Massachusetts Institute of tecnology sobre o futuro do automóvel. Rio de Janeiro: Elsevier, 2004

WOMACK, J. P.; JONES, D.T.. A mentalidade enxuta nas empresas: elimine o desperdício e crie riqueza. Rio de Janeiro: Campus, 1998.

APPENDIX A - MANAGEMENT MODEL HOSHIN (Jackson, 2006)

Made in United States
Orlando, FL
04 November 2026

90200500R00113